Scrunched Cube Explains in 3D What Makes Molecules Solid, Liquid, or Gas

Understanding the 3D Mechanics of the 'States of Matter'
Given the Arno Vigen Scrunched Cube Atomic Model

By Arno Vigen

© 2017/09 E Arno Vigen

Simple Words to Understand . . . Chemistry, Elements, and Bonds

Why does a Nucleus Stay Together If Protons Repel?

- A Nucleus is Just . . . a Magnetic Ring

Why Don't Electrons Fall into the Opposite-Charged Nucleus?

- Electrons are Just . . . Frightened by Nucleus Magnetics

Electron Shell Chemistry Is Just . . . Scrunched Cube Geometry

- Why are electron shells in sets of 2, then 8, then 8 and such? Can we improve Pauli-aufbau?

Scrunched Cube Electron Shell and Bonding Periodic Chart of Elements

- Understanding Molecular Bonding in the Scrunched Cube Atomic Model

Publication: 2017

Scrunched Cube Molecular Bonding

- Understanding Molecular Bonding in the Scrunched Cube Atomic Model

Simple Words to Understand . . . Gravity and Other Forces

Gravity is Just . . . That Electrons are a Little Closer

- Explaining Gravity from the basics of Electromagnetism and Explaining Why Observed Mass Changes

The Five Continuous Fundamental Electromagnetic Forces: Reconnecting Newton into the Chemistry and Particle Physics

- Resolving Strong Force, Weak Force, Bonding Force, Gravity, Mass and $E = mc^2$ via the basics of Electromagnetism as One Continuous Function

Scrunched Cube Explains in 3D What Makes Molecules Solid, Liquid, or Gas

And Why is Gas of Every Element the Same Volume (a mole)?

Publication: 2017

Fixing Einstein's E=mc-Squared

- Integrating that Mass is Magnetics divided by the Electron Shell Radius Volume Explains

How is Electricity and Magnetism Linked?

- Exploring the Fundamental Linkage of Charge and Magnetism

Does Time and Space Really Warp?

- Replacing Electron-Shell Radius for Time-Space Factors in formulas such as the General Theory of Relativity

Simple Words to Understand . . . Personality

Visual Astrology: Fun, Support, Security, and Growth

- Astrology 'signs' archetypes are based upon powerful traits to understand people

Visual Astrology Relationships

- What happens when 'sign' personalities interact

Visual Astrology and Jung

- Astrology 'signs' archetypes actually predict all the Jungian 4 archetypes

Dominant Personality Traits

- Dominant Personality Traits Follow from Four Dimensions, Six Steps and so 24 Subcategories

Simple Words to Understand . . . Communications

Decision Matrix® Writing

- Persuasion is based making arguments at the correct strength in a certain order.

GATESOUP® Writing

- **G**oal, **A**udience, **T**heme, **E**nough **E**lements, **S**upport and the rest

Kedarf® Grammar and Composition Explained

Defining the Parts of Speech, Paragraph Structure and More in Usable Terms

Table of Contents

Scrunched Cube Approach – Atoms, Molecule are 3D; Particles Move as Sets 9

Scrunched Cube States of Matter 10

    Basic AVSC Postulates of Molecules 12

    The Fundamental Forces All From Electrostatic Charge and Nucleomagnetics.. 13

    Electrons, Bonds, are Based upon Open-Paths to the Protons in the Nucleus 18

Solids – Fixed Bonding Directions, Other Non-Bonding Direction Particle Rotations for Heat.. 22

    Solid State Atoms do not Rotate/Orbit.................................... 25

    Electrons in solid settle, and only have minimal Quantum Mechanics distribution 26

    In a Solid State, Electrons Settle and Re-Oriented to Most Compact State 26

Liquid – Repeated Replacing Bonds by Rotational Torque 28

Gas – Rotating so fast that Bonding... 31

    Gas 'State' Takes up a Huge volume 32

Why does Ice Float? Why H2O Has Larger Volume (is less dense) as Solid? 34

Why do Alloy Stainless Steel Bond Differently than Just 26-Fe Iron? 35

    Bonding of 26-Fe Iron alone versus Bonding when Carbon (as in Steel) 37

The Ideal Gas Law AVSC Calculation 41

    Calculating the Boltzmann's Constant 'R' ideal gas by electrostatic charge, nucleomagnetics, and R_{ES}....................... 41

Exception to the Ideal Gas Law under AVSC 42

How can Molecules be Repulsive in Gas State, yet Attractive at Long Distances in Gravity? 43

The Interfaces between States of Matter 49

    The Gas-Solid Interface 49

    The Liquid-Air Interface 51

Scrunched Cube Model of Atom 52

    Nucleus is a Magnetic Chain-Ring 52

    Magnetics for a Ring Creates Perpendicular North-South and 'Bagel' Field Strength Shape 55

Electrons Charge Pull from Nucleus versus Magnetics Force Repulsions Creating a Balancing Point – a Shell 57

Adding Other Electrons Makes the Distances Different After First Two Go North-South Because Magnetic Fields Vary By the Angle 58

Magnetic Poles for First Two of every Shell (1m2 naming versus 1s2) 65

Scrunched Cube Shell 2 and Shell 3 (2c6 naming versus 2p6) .. 66

    From the Magnetic Poles building (1+3+5) x 2 = 9 x 2 = 18 .. 68

    From the Magnetic Poles building (1+3+5+7) x 2 = 16 x 2 = 36 70

Shells Come In Doubles As 2nd Layer Fits in Opening of Lower Layer 71

    From the Magnetic Poles building 1 x 2 = 2 71

    From the Magnetic Poles building (1+3) x 2 = 4 x 2 = 8 71

    From the Magnetic Poles building (1+3+5) x 2 = 9 x 2 = 18 .. 72

    From the Magnetic Poles building (1+3+5+7) x 2 = 16 x 2 = 16 73

Adding Other Electrons Repulsion to Charge-Magnetics Balancing 76

Double Shells Creating Room 2 poles x N squared 76

Transitions Metals –Transitional EndCap Pyramid (4m2+4t6 vs 4d10) 76

Replacing Pauli/Aufbau Logic with 3D Geometry Transition Models 77

Arno Chart of Elements 78

Electromagnetics and the Scrunched Cube 87

Charge-Magnetics Force and Distance 88

Distance and Electron Shells – One Atom 89

Volume Generally Linear to (Z+N) Magnetic Force of Protons and Neutrons 89

Overlap 89

Distance and Molecular Bonding 90

Angles and Electrons Shells – One Atom 91

Angles and Molecular Bonding 92

Larger Atoms No Angle, More Underlying Layer Repulsions 93

Movement of Molecules 94

Gravitational Orbit versus an Electron-Shell Orbit 94

Rotation and Molecules 95

Non-Bonding Cohesion or Adhesion 96

Specific Chemical Processes 97

H2O Water 97

CO2 Carbon Dioxide 98

H2O – Water 118

Plastics – Polycarbonate Chains 127

Salts and Ions 128
Petrochemicals and Energy Release 129
Endnotes 130

Scrunched Cube Approach – Atoms, Molecule are 3D; Particles Move as Sets

All the Arno Vigen Scrunched Cube (AVSC) Atomic Model postulates start with a few basic assumptions that return us to the universal truths of Sir Isaac Newton. We calculate everything based upon Electrostatic Charge and Nucleomagnetics, the two Fundamental Forces. All the interactions and 'states of matter' are determinate by those, and we can and will calculate them.

Even better, we can display these 'state of matter', their properties, and why they happen in vivid 3D pictures and animation.

The big thing for this book is that atoms and molecules are sets where the particles: protons, neutrons, and electrons generally move as a set. It makes a chemistry set that is not just atoms with a block for a bond, but the actual particles, positions and why and how they bond, and what they do in the 'states of matter'. It will show that the particles in solids are also have a solid position. That liquids and gases have 3D visual and clear formulas, from reasons that I can teach to any level student.

Big Hugs,

Let's get going.

Scrunched Cube States of Matter

Universities teach that there are four states of matter:

>Solid
>
>Liquid
>
>Gas
>
>Plasma

Solids are very definite things, you can see that in a rock, or even better for my later examples, a solid crystal, like a diamond with fixed angles that make any person understand that molecules can build structures at specific, consistent angles. Those angles are not just consistent between molecules; the angles actually flow from understandable 3D mechanics of all the particles. Some solids, the crystals have very specific angles between atoms or molecules when combined, and those structures last for long times – millions of years. Solids do not change shapes, solid does not change size, and solid maintain knowable structures and angles – down to the particle level.

On the other hand, the last century of quantum field theory has taught, "one cannot know with any certainty a particle's location and speed; the more you try to understand one of those, the more divergent the other becomes."[i] The Arno Vigen Scrunched Cube Atomic

> Model refutes that concept. Of course, solids have specific location, <u>and their particles are well settled</u>. A is A. That is the only way we can have million year old rocks to climb, and skeletons that keep our bodies from being mush.
>
> That is not to say that quantum mechanics does not have specific uses where it can help determine attributes of electromagnetic vibrations in specific chemicals and chemical reactions. However, quantum mechanics is limited to 'harmonics' of particles that have settling (fixed) positions, but also lots of jostling around in between – like a drum.

Liquids have a relatively, fixed volume, and liquids hold together reasonably well (cohesion). There are no angles. They can move easily into different shapes.

Gases are molecules that are disassociated. One molecules does not bond, even temporarily with another. They fill any size space available expanding greatly from the liquid. The forces holding them are only net-gravity – which is 10^{-34} times less than the electron-proton electrostatic charge force. The a natural volume of a gas is outer space – 10^{+34} more distance between molecules than a liquid.

What engineers calculate more often as gases are disassociated molecules within solid containers, so they bounce off first the wall, and then each other at more than gravity. That leads to the 'enclosed volume ideal gas law':

$$PV=nRT$$

Plasmas are very hot gases in stars, and those take on a different set of reaction properties. There are none here on Earth; and we will not spend much time exploring them.

Basic AVSC Postulates of Molecules

That gets to the basic postulate of the AVSC Atomic Model needed to understand the 'states of matter' in this treatise:

1) Every atom is a linked set. It consists of a) the nucleus (see Book #1 – Why does a Nucleus Stay Together When Protons Repel Each Other), b) its nucleomagnetics field and axis, and c) settling positions of each electron based upon all the forces of all the particles (See Book #3 – Electrons Shells are Really Scrunched Cube Geometry).

2) Electrons settle at the balancing point between a) electrostatic charge attraction towards the protons in the nucleus, and b) nucleomagnetics repulsion for electrons from the set of nucleus particles, protons and neutrons, as constructed, based upon the nucleomagnetics axis angle (See Book #2 – Electrons are Frightened of Nucleomagnetics).

3) The Electrons generally rotates as a set, pushed to stay in the same relative positions as the nucleus and its nucleomagnetics axis rotates. *The major exception to this rule is the Plasma state which I will discuss.*

4) Molecular bonds fit in positions where from another molecules or atom finds an open path to the nucleus proton attractions creates a position (See Book #4 – Scrunched Cube Molecular Bonding). It is a directional attraction and orientation that creates the known bonding angles, distances, and strengths.

The Fundamental Forces All From Electrostatic Charge and Nucleomagnetics in AVSC

All of these are 3D concepts congeal with both deterministic calculations and statistical quantum mechanics applications derived from the AVSC Atomic Model and the associated Fundamental Forces:

Base Fundamental Forces:

- Electrostatic Charge
- Nucleomagnetics

And these combinations of those into the other forces:

- Motomagnetics – the north-south magnetics of movement of or changes to nucleomagnetics field strength.

- Strong nuclear interaction (strong force) holding a nucleus together

- Weak interaction (weak force) keeping electrons in the various Shell and Subshell positions.

- Gravity (See <u>Gravity is Just that Electrons are a Little Closer</u>)

That leads to the main concept of this treatise.

5) The states of matter, are the creating and breaking of bonds; particles settling distances and angles; and their 3D shapes that are calculable with AVSC method of combining electrostatic charge and nucleomagnetics forces.

    In solids, bonds do not break, so both the size, angles, and structures are fixed.

    In liquids, molecules rotate by higher energy, so only temporary bonds form, and those break when replaced with the position of bonding with another molecule, and thereby maintain the size (liquid bond distance) without a permanently fixed structure.

    In gases, bonds do not form between molecules allowing shape changes, but because there is always some bonds at a consistent distance, the volume overall remains consistent.

6) Molecules bonds, adhesion, and cohesion, are depending on the rotational energy of multiple molecules.

7) Solids sit at the particular angles of the Electron Shell in layers relative to the nucleomagnetics axis. The outer shell of electrons do not generally move - other than in the harmonics of quantum mechanics near their settling positions.

   Where those settling positions creates a consistent building structure, you can achieve crystals.

   Other solids lay at angles and distances that by the strong repulsion of outer electrons in their 3D settling position placement on electrons of the combining molecules.

   This makes calculation of alloys much easier in AVSC.

8) Distances for solids, and liquids can get calculated based upon the distances and nucleomagnetics inclination for each particle involved in AVSC.

9) The relatively large distance for Non-Bonding for gases, called the 'enclosed volume ideal gas law', can get calculated based upon AVSC.

10) Plasma occurs when the electron shells themselves loose synch with the nucleus, thereby changing the settling locations, the ability to create bonds, the shells and subshells, and other attributes.

11) The calculation of the average rotational energy, the heat, based upon the AVSC structure, creates the calculation of the melting point, the boiling point for not only Elements and for each compound molecule.

This extra rotational energy builds and twists the molecules until the new state fully changes the 'state of matter' attributes.

12) In liquid, the electrons find open paths, and the molecules short-timing-bond (green line) into those, but there are so many other forces, that the atoms move off. However, there are at *bonding distances*.

13) In gases, both molecules have rotational energy (heat), enough that the attractions when outer electrons in one molecule can find a temporary open path towards the nucleus protons. Those get overwhelmed, until the time in open-path (green) attractions, balance the time in electron-between repulsion (blue).

14) The shape of the electron outer subshells, and how then constructed creates tendencies for other properties like magnetics, electrical resistance, and more.

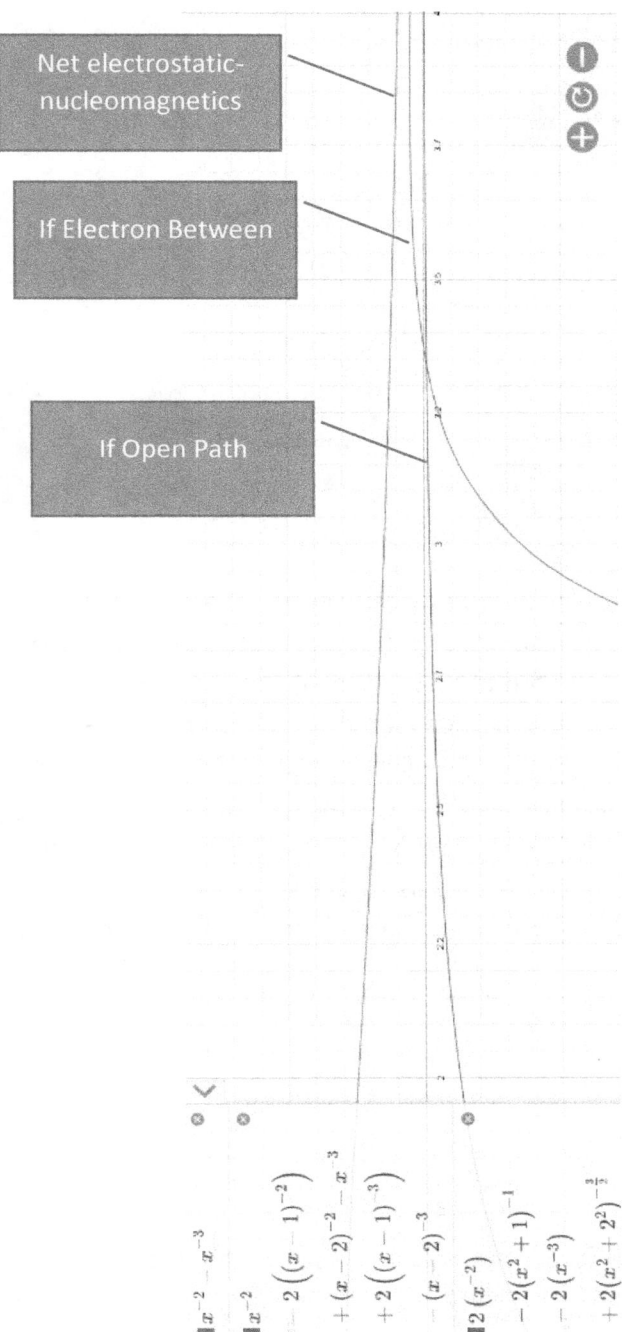

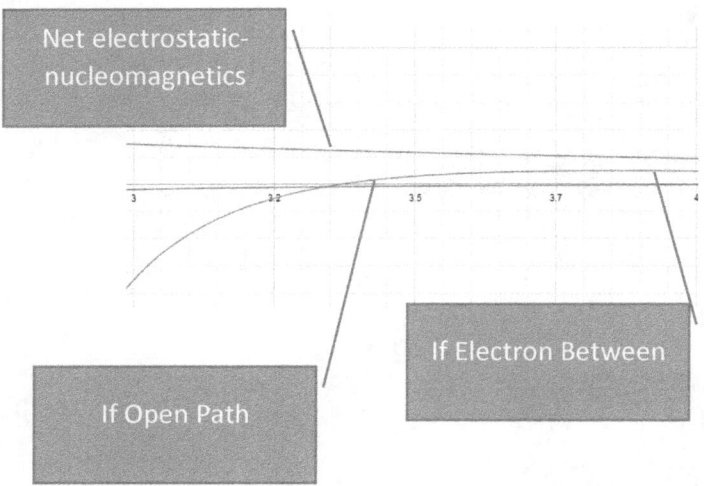

The graph shows the standard, single-particle electrostatic-nucleomagnetics (red) force which decreases towards zero by 1/distance-squared at a distance. It then shows the system for multiple particles along an open path to the nucleus between two single-atom molecules (green) which increases towards zero eventually by 1/distance-squared. Finally, the graph shows a path blocked by external electrons which actually becomes repulsive decreasing towards zero from a repulsive (positive) force at distances get greater.

Bonds are Based upon Open-Paths of Outer Electrons to the Protons in the Nucleus of the Other Atom/Molecule

The open-path force is attractive – approaching zero from the negative side; the electron-between force is repulsive –

approaching zero from the positive side. As the position changes, the net bonding forces keep changing from positive (repulsive - yellow) and negative attractive – green).

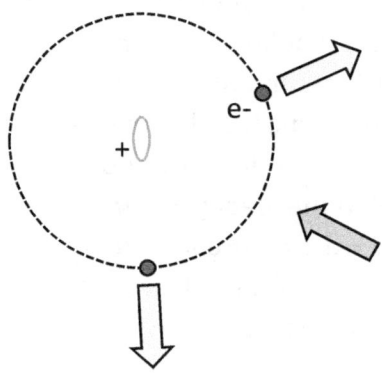

Further, the electron-repulsion decreases faster, by the (x-2)^2 factor, such that those two eventually have the exact opposite strength at a particular distance which defines 'R' in the ideal gas law.

The force become repulsive after someplace after 130x the average radius of the molecules.

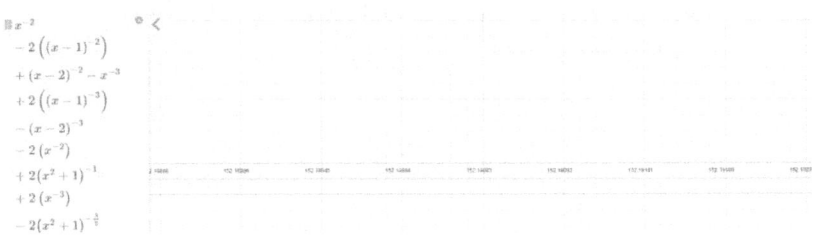

15) That balancing point follow PV=nRT, the ideal gas law. The product of volume times pressure directly related by 'R' to the product temperature times the number of gas molecules.

16) You will see that in 3D as the treatise proceeds. There is a point a) with multiple outer electrons repulsions, and b) enough rotation, so the outer electron exposed always is repulsive. Once that happens, the other rotation creates a larger repulsion equal to the ideal gas law. In these situations, that all repulsive pressure add such that the molecules can stay generally about 140 radius (R_{ES}) apart if in a closed container.

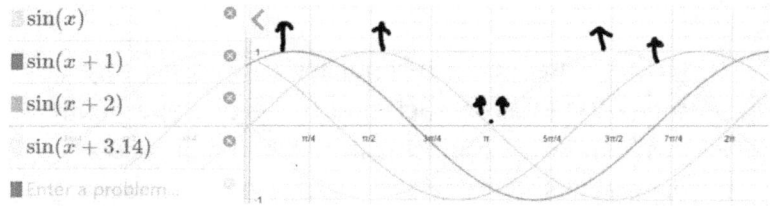

But if the energy is not enough, a portion of the rotation is attractive (yellow), and the two molecules so aling, even weakly into a liquid bond. At that point, the molecules can stay generally about 3-4 radius (R_{ES}) apart.

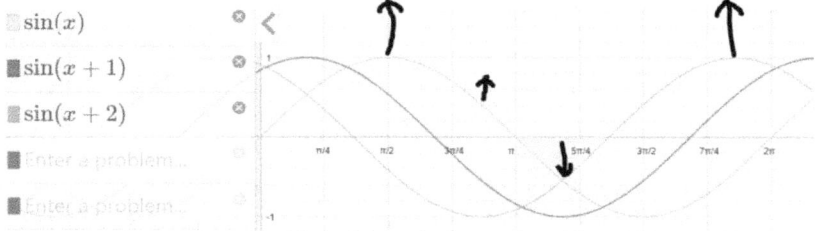

17) You will understand that in mathematics as the treatise proceeds.

18) After the ideal gas law distance, if not contained in a closed space, the repulsion continues to decrease such that the attraction is greater, becoming Newton's Gravity force 'G' over time and sufficient molecules – whether solid, liquid or gas as the distant object.

19) How does it move from repulsive to attractive is the nature of the 1/distance cubed eventually becoming unimportant until the comparison is the integral of the electrons spaced in every direction, reasonably evenly, such that the integral of 1/distance-squared front and back, is slightly less than the attraction of the nucleus 1/distance-squared for its protons.

20) Finally, the extra energy of heat gets taken as rotation of particles along around axis of non-bonding such that we can get states of matter' to remain for a range of temperatures.

Solids – Fixed Bonding Directions, Other Non-Bonding Direction Particle Rotations for Heat

In solids, the fixed Bonding position is based upon a large gap between the number of electrons of the outer set of subshells for an Element versus the full shell configuration.

As background, full shells want to build in two hemispheres according to the nucleomagnetics force and axis. The full shells are gases because all the easy paths are filled. There is not inclination angle/position for permanent bonds or temporary bonds. Those full-shell, Nobel Gas Elements happened to have the number of electrons of the full shells.

	Number of Hemispheres	Number of Electrons in outer	Total Outer Electrons
Shell-1	2x	1-squared = 1^2 = 1	2
Shell-2	2x	2-squared = 2^2 = 4	8
Shell-3	2x	2-squared = 2^2 = 4	8
Shell-4	2x	3-squared = 3^2 = 9	18
Shell-5	2x	3-squared = 3^2 = 9	18
Shell-6	2x	4-squared = 4^2 = 16	32
Shell-7	2x	4-squared = 4^2 = 16	32

These I have builds in 3D SolidWorks which provides structure that we can engineer into bonds and chemical reactions.

In the below, I have a 36-Kr, Krypton noble gas electrons shell structure which is very full. Looking from either the equator (Figure 1) or the nucleomagnetics polar axis (Figure 2).

Equator View — These block any >15 degree open-path

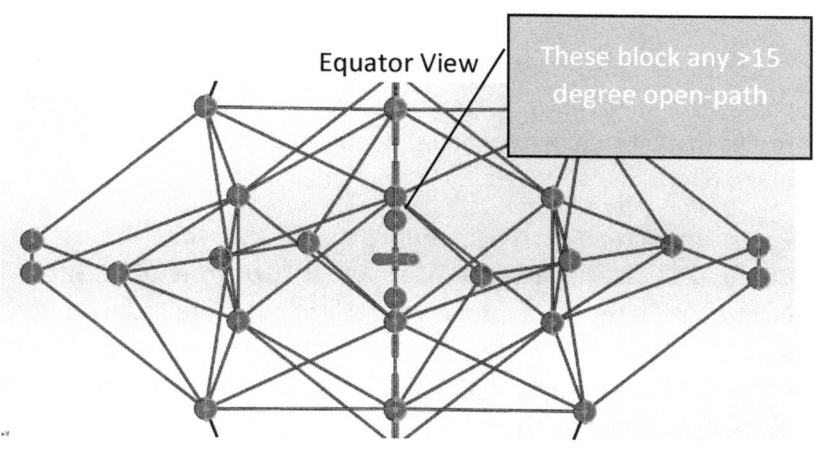

Polar View

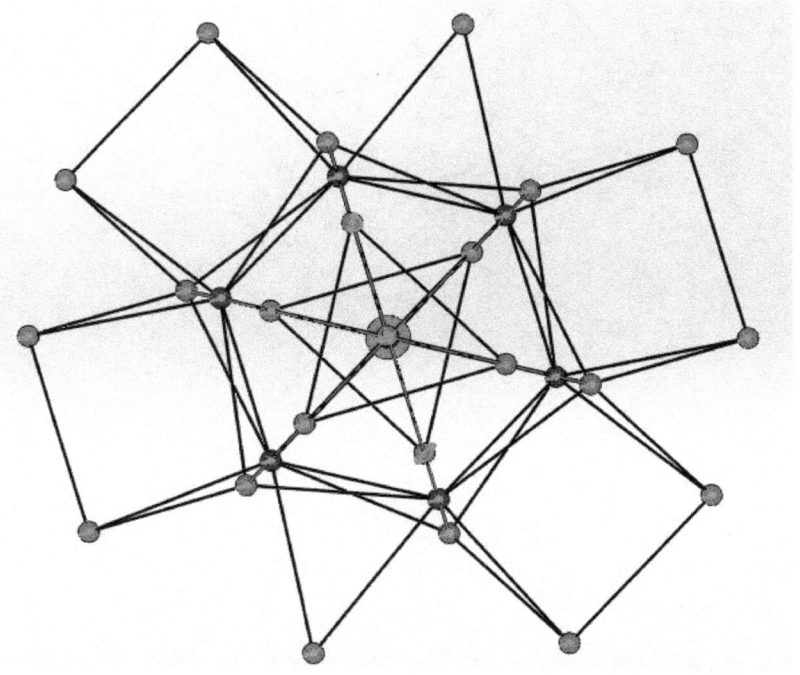

Obviously, actual molecules do not have the 3D CAD/CAM connectors that I used. That connection is really the balancing of forces that vibrate in harmonics (deeply described by quantum mechanics).

For 'states of matter', the important issue is that you need, at 20 Celsius/293 Kelvin, a channel of about 15-20 degrees of the electron repulses the approaching molecules before it can bond as a solid or liquid. As you can see, the noble gas does not have such a channel from these two directions. Every direction has electrons blocking the open-path.

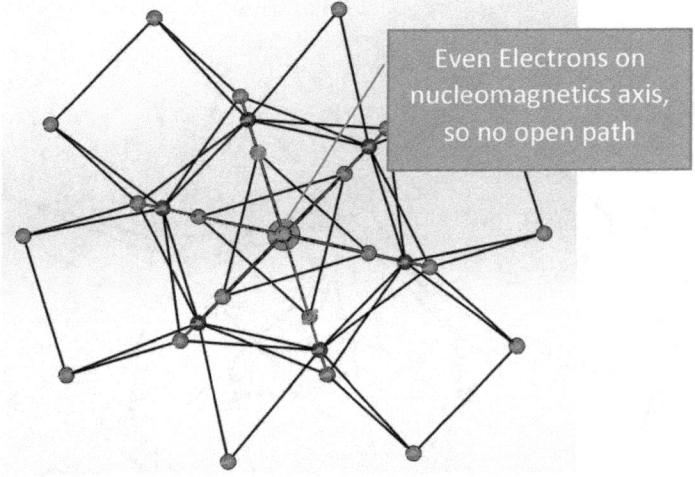

Even Electrons on nucleomagnetics axis, so no open path

That is why these, at all but near absolute zero, noble gases will not create bonds, and thereby operate in gas 'state'. There is not open path for a bond.

For another element, where the electron count is less, the bonded atom orients with its electron filling that position.

Water (H2O)

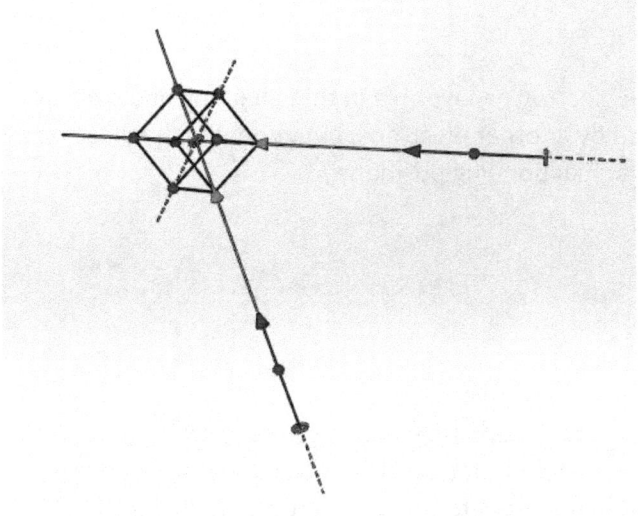

Now, that other atoms cannot fit all the way in; it is more than just a single electron. The bonded atom has other protons and electrons which both create forces such that the position is about 3-4 times the shell distance for the new set.

Solid State Atoms do not Rotate/Orbit

If electrons really always rotate, or exist/non-exist a statistical field, then a crystal could not exist. Once in a solid, the bonding electron location is within a well-defined channel at a nucleomagnetics inclination/longitude angle and latitude/azimuth angle.

Electrons in solid settle, and only have minimal Quantum Mechanics distribution

Yes, the electron can wobble in that channel, and even get replaced by another electrons, but generally the electron fit into and fill actual bonding positions.

In a Solid State, Electrons Settle and Re-Oriented to Most Compact State

There are two interesting steps to a solid. One is that very weak bonds occur where the outer electrons see the lowest point, and a relatively open path to the other atom nucleus.

The interesting thing is that the other electrons in both molecules repel each other. That creates a consistent volume for each molecules, and shape for electrons. The other electrons create a locking force. One electron is pulled toward the open-path to the nucleus, but other electrons, from both atoms, repel each other making the bond at a specific angle.

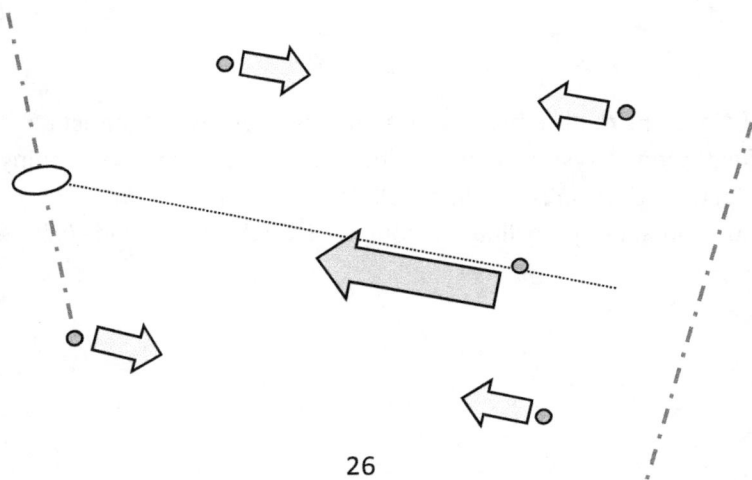

This creates a lock-in of the angle. This is why the angles in solids are constant. While we have a channel 15-20 degrees open to bond, once the bond is formed, the other electrons push that electrons to bond at a narrow range to reduce the repulsion at the sides.

If the electron wants to drive out of the open-path (angle), the nearest electron repulsion grows to push it back into bonding position.

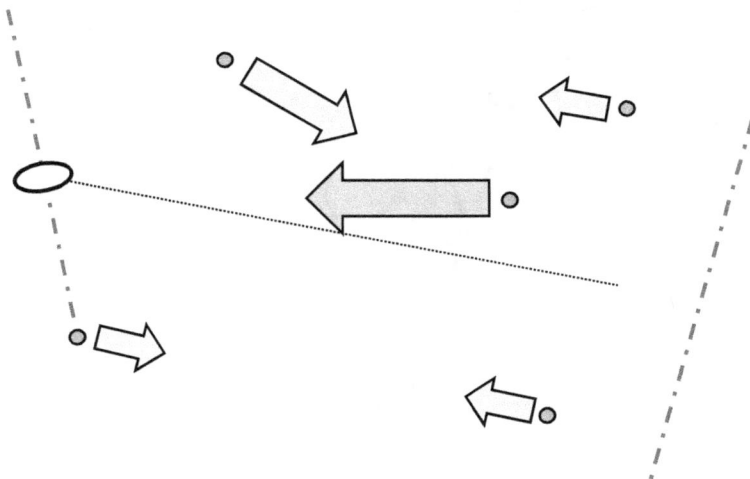

Liquid – Repeated Replacing Bonds by Rotational Torque

However, look at the water (H2O) molecule again. There are lots of directions that do have a direction with an open path from another atom or molecule towards the nucleus (+).

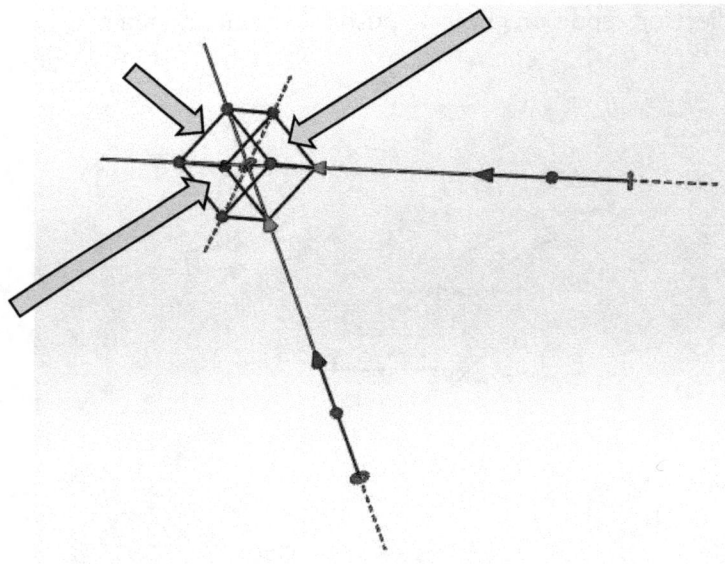

The angle of the corners of the Scrunched Cube for the Oxygen atom is 72 which makes the opening have 36 degrees closer to electrons and 36 degrees open to the nucleus. That is big enough (>20 degrees) for solid bonding – either as a solid or liquid depending on other factors (temperature).

If the molecule were not moving, it will bond as a solid; water does up to 273 degrees Kelvin. It will operate as liquid up to 373 degrees Kelvin.

The secret of liquids is that molecules is bumped around. It has spin along multiple axis. That means that open paths keep rotating (a sine wave), and most of the time an electron gets in the way before a permanent bond can form.

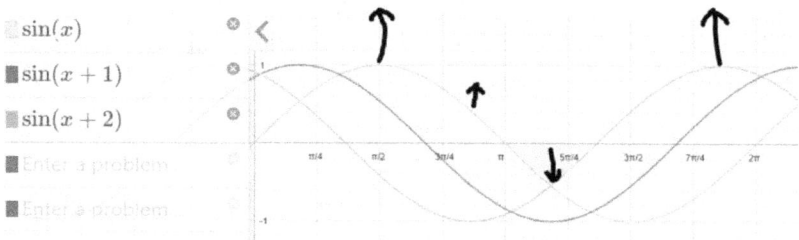

In a liquid, another molecule finds an open path, but there is enough rotational energy (heat/temperature) such that the atom rotate with enough force to break that bond, but has likely created another bond it in its place. The open path (green) moves to blocked (yellow) as an electron rotates in between.

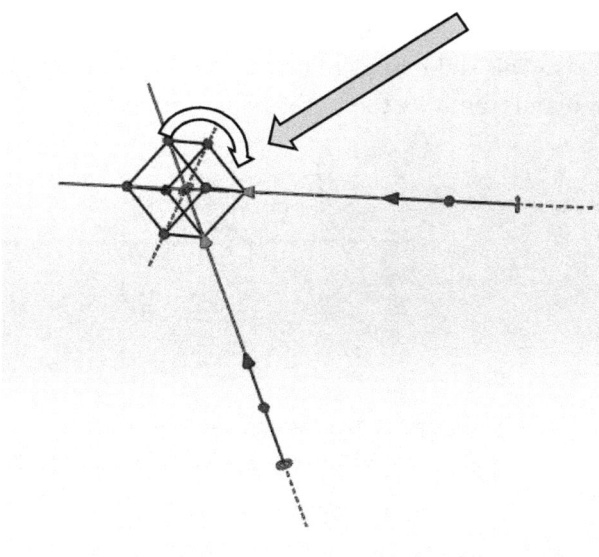

In liquids, the bonds do not have the strength versus the torque of that level of rotation (heat/temperature) that breaks them. However, they have enough strength for lots of temporary bonds.

In solids, that means that atoms will re-orient until they have the most bonds, and the smallest space. In physics, that is the lowest entropy. In solid, you end of with distances at the minimum of these graph. In liquid, you get into the bottom, but then get bumped out.

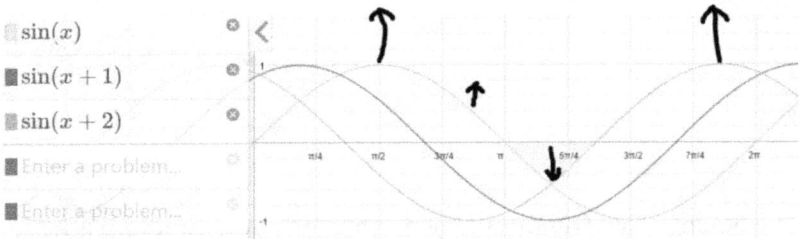

However, if you get to a place where not positions have attraction (negatives), there is the magic jump to the maximum of the circuit. Nothing is pulling them together, so they want to push out as much as possible (the maximum).

That is why we see such an expansion of volume when liquids move to gas. This is the basis of an air conditioner heat pump. The gas expands, and the temperature goes down.

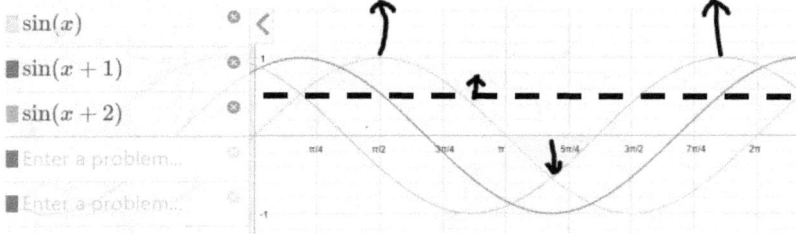

In the gas state, the phases are moving so fast, that the atom cannot bond, it just settles at the outer distance (the dashed line)!

Gas – Rotating so fast that Bonding

The calculation when the electrons are out of the way, the net force is attractive (physics-negative).

		Position
	Molecule 1:	
(Z,N)1	Nucleus	0
e1	Electrons	1
	Molecule 2:	
(Z,N)2	Nucleus	x
e2	Electrons	x (at side)

		Position1	Position2	Distance	Electrostatic					Nucleomagnetics		
(Z,N)1	(Z,N)2	0	x	x	X-Repuls				x-Attraction			
(Z,N)1	e2	0	x (at side)	root(1^2+x^2)	root(1^2+x^2-Attraction)					root(1^2+x^2-Repulsion)		
e1	(Z,N)2	0 (at side)	x	root(1^2+x^2)		root(1^2+x^2-Attraction)					root(1^2+x^2-Repulsion)	
e1	e2	0 (at side)	x (at side)	x			x Repul					x-2 Attraction
												0.12500000
		x	Y	x^(1/2)	(x-1)^(1/2)	(x-1)^(1/2)	(x-2)^(1/2)	x^(1/3)	(x-1)^(1/3)	(x-1)^(1/3)	(x-2)^(1/3)	
		2	-0.22111456	0.25000000	(0.20000000)	(0.20000000)	(0.25000000)		0.08944272	0.08944272		
		2.5	-0.17340947	0.16000000	(0.13793103)	(0.13793103)	(0.16000000)		0.05122630	0.05122630		
		3	-0.13675445	0.11111111	(0.10000000)	(0.10000000)	(0.11111111)		0.03162278	0.03162278		
		3.1	-0.13063096	0.10405827	(0.09425071)	(0.09425071)	(0.10405827)		0.02893523	0.02893523		
		3.2	-0.1248621	0.09765625	(0.08896797)	(0.08896797)	(0.09765625)		0.02653692	0.02653692		
		3.3	-0.11942685	0.09182736	(0.08410429)	(0.08410429)	(0.09182736)		0.02439087	0.02439087		
		3.4	-0.11430473	0.08650519	(0.07961783)	(0.07961783)	(0.08650519)		0.02246547	0.02246547		
		3.5	-0.10947606	0.08163265	(0.07547170)	(0.07547170)	(0.08163265)		0.02073367	0.02073367		
		3.7	-0.10062505	0.07304602	(0.06807352)	(0.06807352)	(0.07304602)		0.01776100	0.01776100		
		3.8	-0.09658823	0.06925208	(0.06476684)	(0.06476684)	(0.06925208)		0.01648273	0.01648273		
		3.9	-0.09273592	0.06574622	(0.06169031)	(0.06169031)	(0.06574622)		0.01532235	0.01532235		
		4	-0.08911346	0.06250000	(0.05882353)	(0.05882353)	(0.06250000)		0.01426680	0.01426680		
		5	-0.06183722	0.04000000	(0.03846154)	(0.03846154)	(0.04000000)		0.00754293	0.00754293		
		6	-0.04516762	0.02777778	(0.02702703)	(0.02702703)	(0.02777778)		0.00444322	0.00444322		

The calculation when the electrons are in the path, the net force is repulsive (physics-positive).

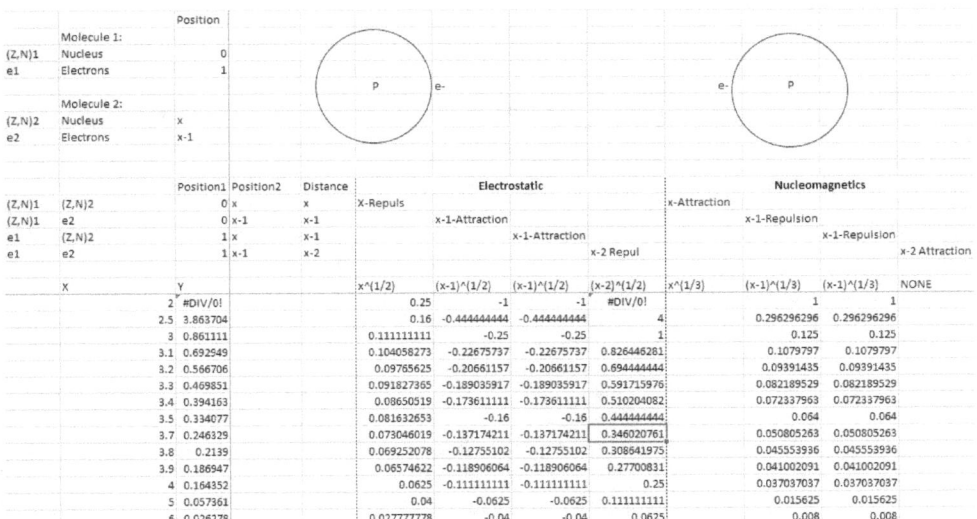

		Position
	Molecule 1:	
(Z,N)1	Nucleus	0
e1	Electrons	1
	Molecule 2:	
(Z,N)2	Nucleus	x
e2	Electrons	x-1

		Position1	Position2	Distance	Electrostatic					Nucleomagnetics		
(Z,N)1	(Z,N)2	0	x	x	X-Repuls				x-Attraction			
(Z,N)1	e2	0	x-1	x-1		x-1-Attraction				x-1-Repulsion		
e1	(Z,N)2	1	x	x-1			x-1-Attraction				x-1-Repulsion	
e1	e2	1	x-1	x-2				x-2 Repul				x-2 Attraction
		x	Y	x^(1/2)	(x-1)^(1/2)	(x-1)^(1/2)	(x-2)^(1/2)	x^(1/3)	(x-1)^(1/3)	(x-1)^(1/3)	NONE	
		2	#DIV/0!	0.25	-1	-1	#DIV/0!		1	1		
		2.5	3.863704	0.16	-0.444444444	-0.444444444	4		0.296296296	0.296296296		
		3	0.861111	0.111111111	-0.25	-0.25	1		0.125	0.125		
		3.1	0.692949	0.104058273	-0.22675737	-0.22675737	0.826446281		0.1079797	0.1079797		
		3.2	0.566706	0.09765625	-0.20661157	-0.20661157	0.694444444		0.09391435	0.09391435		
		3.3	0.469851	0.091827365	-0.189035917	-0.189035917	0.591715976		0.082189529	0.082189529		
		3.4	0.394163	0.08650519	-0.173611111	-0.173611111	0.510204082		0.072337963	0.072337963		
		3.5	0.334077	0.081632653	-0.16	-0.16	0.444444444		0.064	0.064		
		3.7	0.246329	0.073046019	-0.137174211	-0.137174211	0.346020761		0.050805263	0.050805263		
		3.8	0.2139	0.069252078	-0.12755102	-0.12755102	0.308641975		0.045553936	0.045553936		
		3.9	0.186947	0.06574622	-0.118906064	-0.118906064	0.27700831		0.041002091	0.041002091		
		4	0.164352	0.0625	-0.111111111	-0.111111111	0.25		0.037037037	0.037037037		
		5	0.057361	0.04	-0.0625	-0.0625	0.111111111		0.015625	0.015625		
		6	0.026278	0.027777778	-0.04	-0.04	0.0625		0.008	0.008		

Gas 'State' Takes up a Huge volume

That means that atoms will re-orient until they have the most bonds, and the smallest space. In physics, that is the lowest entropy.

In solids, you end of with distances at the minimum of these graph.

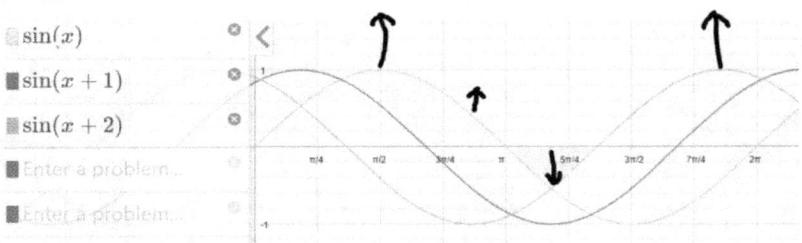

In liquid, you end of with distances at something like the average these graph.

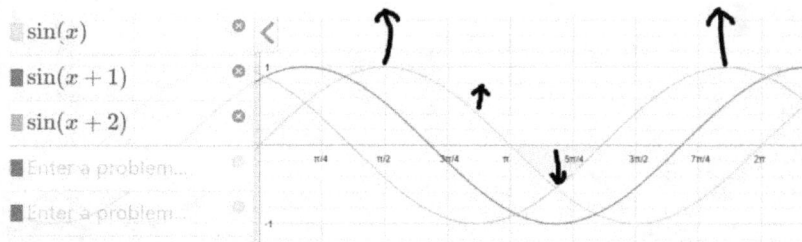

Remember that nothing holds particles. Unless other forces (torque from rotation energy/heat), the particles will flow to the low point.

However, if you get to a place where not positions have attraction (negatives), there is the magic jump to the maximum of the circuit. Nothing is pulling them together, so they want to push out as much as possible (the maximum).

That is why we see such an expansion of volume when liquids move to gas. TNT is the process where each solid molecule breaks into three gas Nitrogen's and Oxygen's, each gas molecule creating its own huge volume – an explosion thrust at 1,000 meters per second.

This is the basis of an air conditioner heat pump. Compress to liquid takes heat from in building to out where it cools, then on trip back the gas expands, and the temperature goes down.

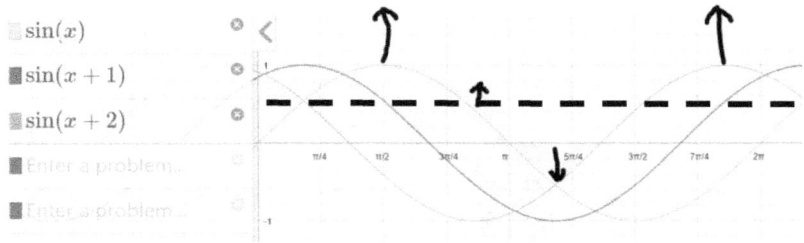

In the gas state, the phases are moving so fast, that the atom cannot bond, it just settles at the outer distance (the dashed line)!

Why does Ice Float? Why H2O Has Larger Volume (is less dense) as Solid?

Normally, a liquid is larger than its' solid. As it rotates, it is pushing other atoms to bond further, or not bond as deeply. In that sense, the normal process is that the density of the solid is more than the density of the same molecules as a liquid (at higher temperature).

However, water is a known exception. We know that water floats – like in the iceberg that sank the Titanic. Water actually gets closer (shrinks) at the higher temperature liquid state; that is unusual.

Let's look at the picture again, the solid bonds are often to the Oxygen, but the solid bonds are with the Hydrogen, which for the H2O molecule is actually further out:

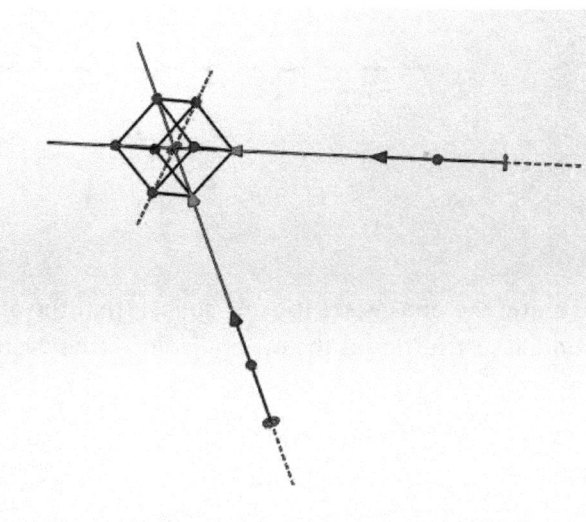

The secret is that water has Hydrogen in bonds. Those O1-H Hydrogen atoms do not settle with a) electrons on the outside. That means that, in the solid, the next bond starts at the proton at the end of the Hydrogen.

At Hydrogen Proton-Outwards to another H2O atom is solid. There are no surrounding electrons to add repulsion. As such, the bonds in solid are from the H2O Hydrogens, to the other H2O Oxygens.

Yet, for the liquid, temporary bonds form with the Oxygen, even Oxygen to Oxygen, which are about the same distances as the H2O Hydrogen. As such, the liquid bonds are closer (O::O) versus (O::H) bonds of solids. The liquid can slip inside and have more density.

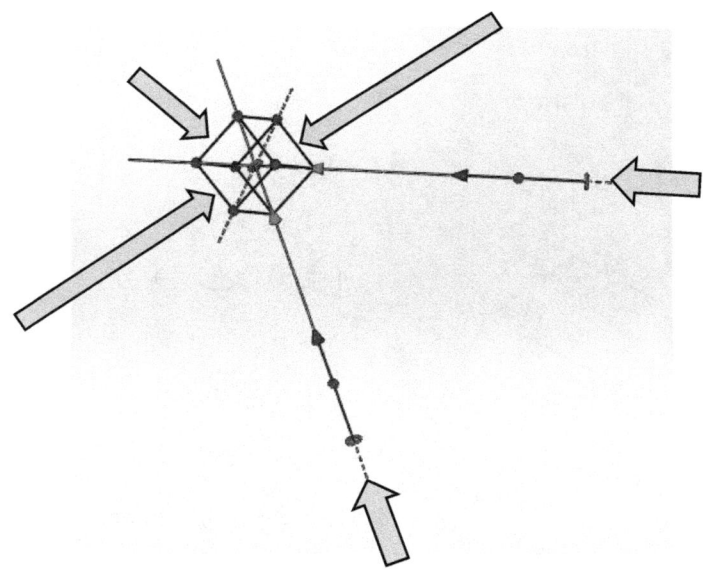

Why do Alloy Stainless Steel Bond Differently than Just 26-Fe Iron?

We are all familiar with iron and steel.

In AVSC, a 26-Fe Iron atom is an oblong zone (outer shell) of electron-repulsion. It has the inner scrunched cube, but the 4th shell is incomplete, consisting of eight (8) electrons. Those are in an endcap at each nucleomagnetics pole.

However, there are not eight exterior electrons, but fourteen (14), and to a slight degree another six from Subshell-3c.

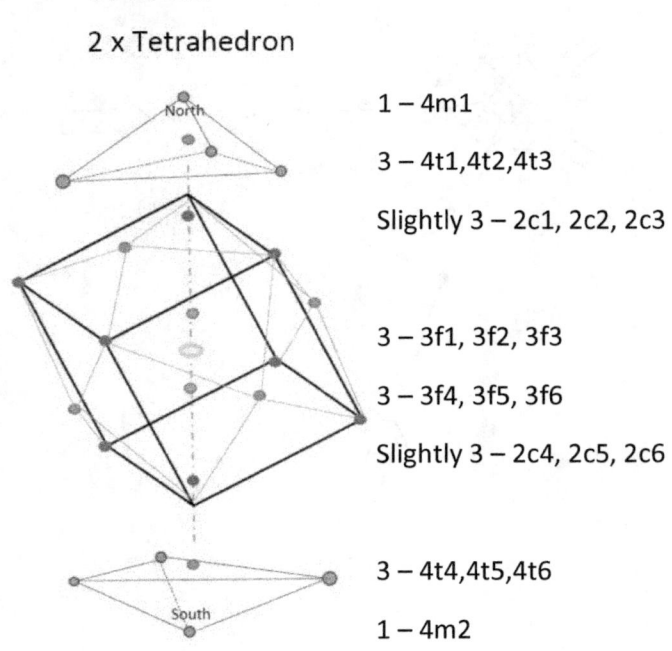

26-Fe Iron

2 x Tetrahedron

1 – 4m1

3 – 4t1, 4t2, 4t3

Slightly 3 – 2c1, 2c2, 2c3

3 – 3f1, 3f2, 3f3

3 – 3f4, 3f5, 3f6

Slightly 3 – 2c4, 2c5, 2c6

3 – 4t4, 4t5, 4t6

1 – 4m2

This makes the electromagnetic spectrum a huge set of combinations.

Once, you add additional electrons in 27-Co Xo Cobalt or 29-Cu Copper, the Subshell-4u blocks any bonding or radio emissions form the 2f6 and 3c6 electrons. You can see this in the spectrum changes to a few lines versus dozens. 26-Fe Iron is the last in the red circle with lots of spectrum lines. 29-Cu Copper is in the yellow circle, with just a few spectrum lines.

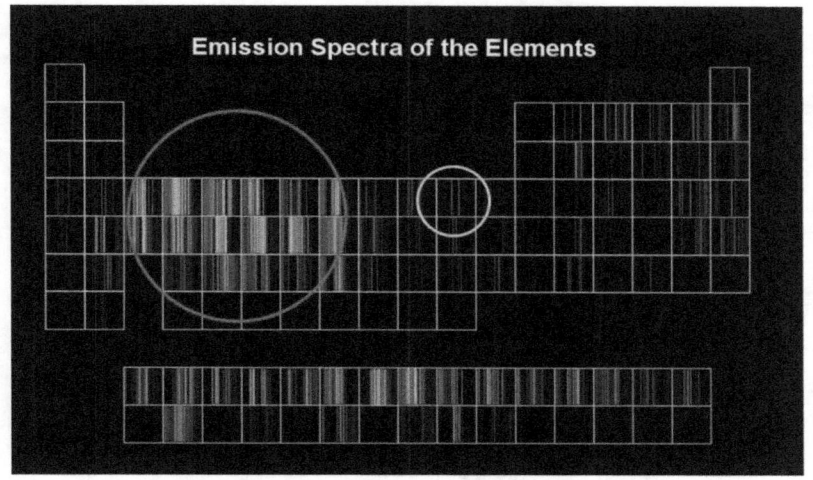

Bonding of 26-Fe Iron alone versus Bonding when Carbon (as in Steel)

26-Fe Iron alone can has this oblong shape, and in solids, they outer electrons find a variety of places to settle. That leaves lots of places sticking out, and available to rust.

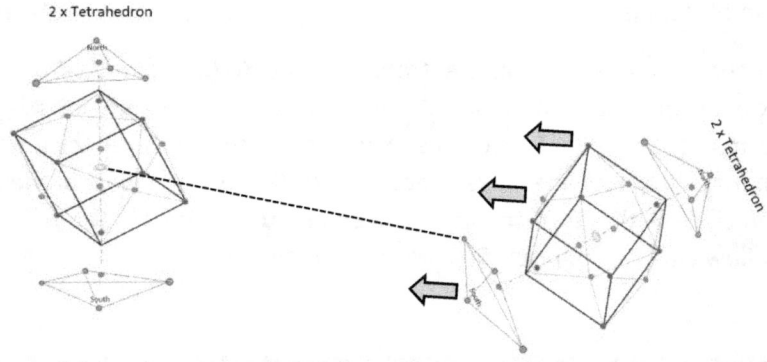

However, when the correct percentage (about 5%) of Carbon is added, the alloy bonding creates a structure that is more consistent in orientation, more dense, and less reactive.

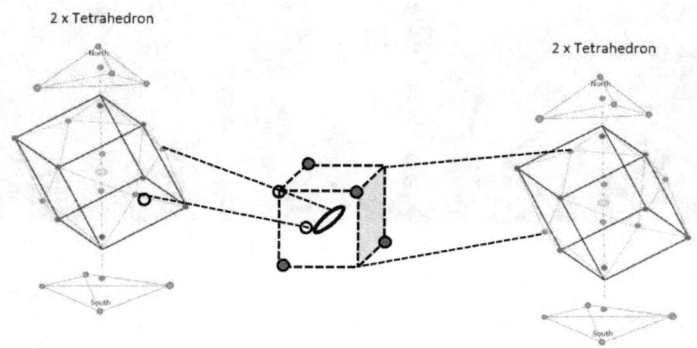

The outer electrons of the 26-Fe Iron atom finds open paths (corners of the Carbon), where they can bond the two atoms. There are only two (2) repulsing electrons versus seven (7). Two of the 26-Fe Iron outer electrons have an open path in the wide open Carbon which does not occur in the all 26-Fe Iron solid.

That Carbon bond is stronger than Oxygen or water, so this Steel (iron with 5% carbon) does not rust significantly when wet or exposed to air compared to the Iron only structure.

Further, this gives Iron structure to make it strong versus the every direction weak pure Iron bonding.

The key element of bond strength is the number and positioning of electrons that interact with exterior electrons of the other bonding atom. The repulsion of a Carbon comes from four (4) other electrons.

Carbon

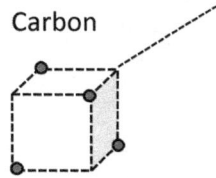

The repulsion component for an Oxygen bond comes from six (6) other electrons.

Oxygen

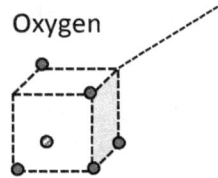

The repulsion for 26-Fe Iron bonding comes from 14 outer electrons, so its repulsion is greatest, and bond strength the lowest.

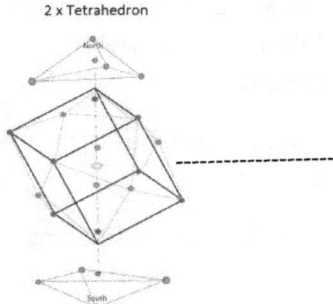
2 x Tetrahedron

Element	Repulsion from	Bond Strength
06-C Carbon (Receiving)	4 Electrons	Highest
08-O Oxygen	6 Electrons	Medium
26-Fe Iron (Receiving)	14 Electrons	Lowest

If only Iron, the Oxygen bonding is better than bonding itself, so iron likes to rust. The bond strength of the iron when already bond Carbon (in steel) makes the likelihood of Oxygen bonds reduce substantially. The Carbon bond is tighter making the Oxygen always lower, and unlikely. It will always preference staying with Carbon over Oxygen.

If the Iron is with Iron, the Oxygen actually replaces the Iron as the primary bond.

This creates the manufacturing balancing act. No carbons, and the bonds are weak, and not aligned. With a few Carbon atoms, two Carbons (and each of those next layer of eight) directly bond or orient Iron atoms; so at about 5% (1/18) the bonding is most effective. But if you get too many, the Carbon will we at every opening, and actually pull Oxygens into the molecule and cause bond weakening and/or rust.

The Enclosed Volume Ideal Gas Law AVSC Calculation

Once a molecule is rotating fast enough, it is gets to a gas 'state' where the rotating molecules are always pushing other nearby atoms at standard bonding distances.

This creates a structure where the molecules have a area closes that is repulsive, yet at great distances, it is attractive, with a balancing point – the ideal gas law.

PV = nRT

P = pressure

V = volume

n = number of moles of in the gas (number of atoms without regard to the particular Element or compound)

R = Boltzmann's Constant

T = Temperature

Calculating the Boltzmann's Constant 'R' ideal gas by electrostatic charge, nucleomagnetics, and R_{ES}

Exception to the Ideal Gas Law under AVSC

There are exception to the ideal gas law. Places where the rotating of the molecule does not get all electron repulsions.

One is water (H2O). Let's look at the Hydrogens now. The 01-H Hydrogen's electrons settle facing inward to fill the open path two positions 2f5 and 2f6 toward the Oxygen.

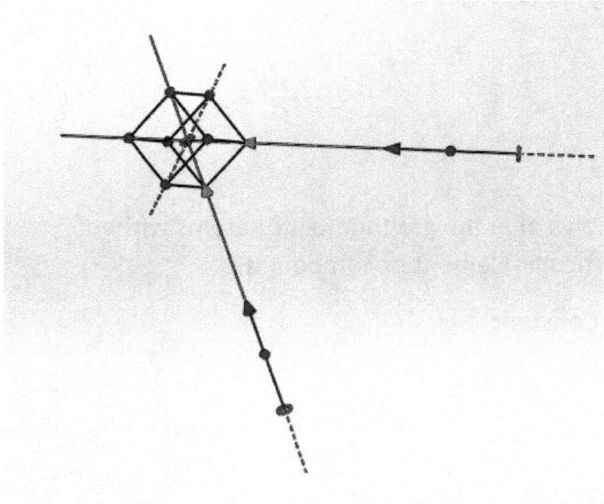

However, this is another huge difference, the Hydrogen settle proton-outward. That means not matter how fast you rotate (how much heat your apply), a water will always have this one loop of positive charge – limiting the repulsion calculation. As such, water does <u>not</u> follow the ideal gas law.

The same applies for hydrocarbons.

How can Molecules be Repulsive in Gas State, yet Attractive at Long Distances in Gravity?

Great question about AVSC. The basic challenge is that we have molecular structures are that are very balanced. Electron counts equal Proton counts. As a results, the first level of thinking would say nothing bonds because everything is balance equally. Of course, that is not true.

Electrons have settling positions, and that means some directions, where there is an open path, become attractive, yet other direction, where electrons settle, become repulsive.

The AVSC basic gravity theory shows that if you have lots of these molecular systems, not directly interacting with each other, then the electrons set at every level, and the

In gravity, there is a tiny next much smaller (10^{-34}x) the electrostatic charge force. It really is right on the border.

In the states of matter, the force is sometimes attractive (permanent bonding in solids, temporary bonding in liquids) and sometimes repulsive (in highly rotating gas molecules). So, again, the forces generally balance, but there are situations where one is greater than the other based upon the circumstances.

However, that makes for a flip from repulsive in the state, and then magically it flips to attractive at a greater distance for gravity without an additional particle. That means that we have:

Situation	Distance	Force Strength
Solid	Touching	Attractive
Liquid	In a closed container	Attractive
Gas	In a closed container	Repulsive ** why?
Gravity	Distant objects	Attractive

The different is the situation and the rotation of the molecular systems.

In a solid 'state of matter', the molecular system is not rotating relative so, two molecules can bond' into the lowest value, the physics-negative attractive. It can rotate and stay in the yellow area.

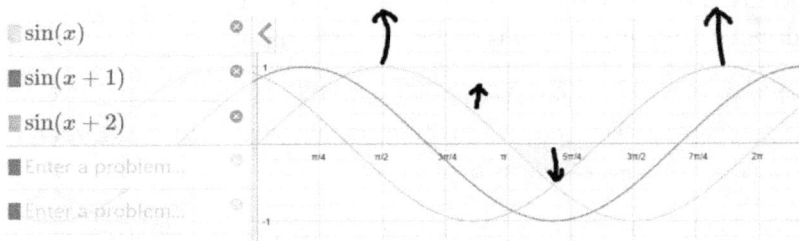

Similarly, the liquid 'state of matter', the molecules can find that low point, and bind, but the rotational energy (temperature) then often breaks those bonds. However, that rotation then puts another molecule to do that same process of bond/break, keeping the multi-molecule system in the yellow, physics-negative, attractive state.

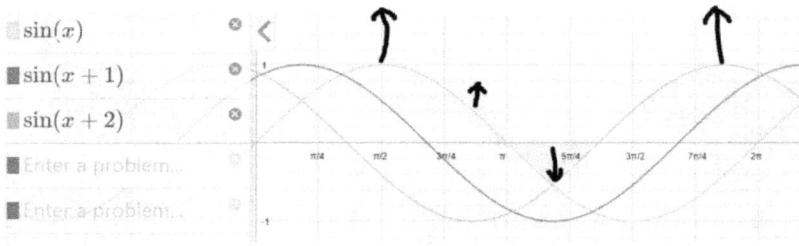

For gravity, the distant particles are not interfacing, and the mathematics that the front 1/distance-squared attract more than the back 1/distance-squared particles repel. That makes a tiny net, gravity.

The following explains the general math principle that a molecular system, beyond a gravitational orbit of large masses, where the nucleus and electrons are held by the balancing of electrostatic charge and nucleomagnetics, the system must generate a tiny attraction calculation.

A simplified example will explain why the nucleus-shell structure with electrons in 'orbit' generates a 'tiny' net force at 1/-distance squared:

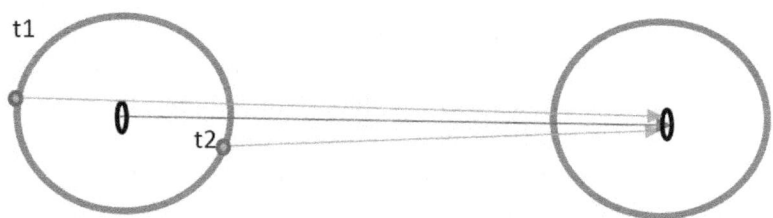

For simplicity[ii]: compare the force which acts using the distance-squared (as is the case for electromagnetic Charge and gravity) when the nucleus sits at a simple 10, say meters, (dropping all zero exponents/E-XX) and the electron moves in an orbit of 1.

That makes the nucleus each sit at 10 apart, and the electrons orbits at a distance from 11 at time1 and 9 at time2. The system moves back and forth 11 then 9, then 11, then 9.

Interaction / Time	Force / Distance-Squared	Calculation	Subtotal (in Force units)	%
Proton<>Proton				
Time 1	$-1.0 / 10^2$	0.01000		
Time 2	$-1.0 / 10^2$	0.01000		
- Total			-0.02000	100%
Electron <>Proton				
Time 1	$1.0 / 9^2$	0.01234		
Time 2	$1.0 / 11^2$	0.00826		
- Total			+0.02060	103%
Grand Total			+0.00060	3%

So, whenever there is a nucleus-shell configuration, charge *must* create some net-over-time (as they orbit) force towards distant objects. It is not 3% because my example numbers do not reflect the actual shell-distance difference ratio, but there is some 'tiny' extra-net charge force that has not been included in previously calculations trying to determine gravity.

The counter-argument that electrons and nucleus rotate around each other, in a gravitational orbit is address in my prior book – <u>Gravity is Just that Electrons are a Little Closer</u>.

However, the original table is wrong. Gas is also attractive.

If you pour gasoline into an open box, the gas will tend to stay together. In fact, the Earth's atmosphere is a great example. Here is an open container, the Earth, but the gas does stay here. Venus and Jupiter have miles and miles of gases that do not repel each other. That table is false.

In outer space, we have seen the NASA astronaut video that water will form its own perfect floating drop. Water is attractive also when alone.

The problem with the table is that we are using a closed container, and the pushback of the solids of the walls **is** greater than gravity.

Situation	Distance	Force Strength
Solid	Near each other in open space	Attractive
Liquid	Near each other in open space	Attractive
Liquid	In a closed container	Attractive
Gas	Near each other in open space	Attractive
Gas	In a closed container	Repulsive ** because the wall's pressure translates to all of the space greater than gravity
Gravity	Distant objects	Attractive

With nothing else, the basic same Element forces, following the AVSC electrostatic charge-nucleomagnetics curve.

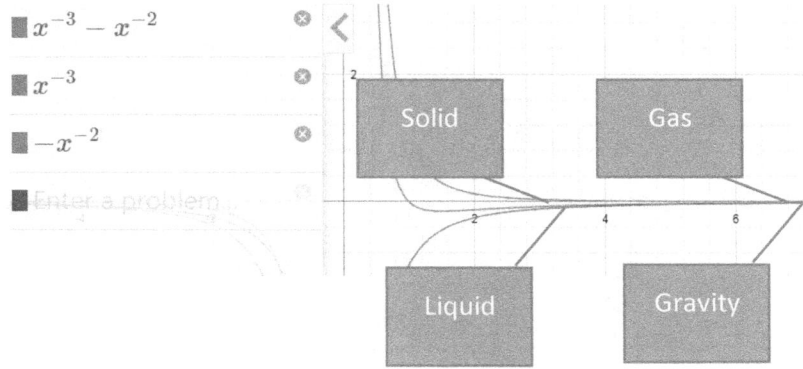

47

The force is always attractive. The distance for solids is tightest, liquids (except H2O discussed separately) next, gas is further out the curve at 6x, 20x, 133x the radius of the electrons shell (R_{ES}). Finally, gravity is everything thousands or millions away with the reduction of $4/3\pi R_{ES}$ to calculate the integral of front and back electrons over time.

However, that would mean that we move to the average – which is zero or slightly physics-negative attractive.

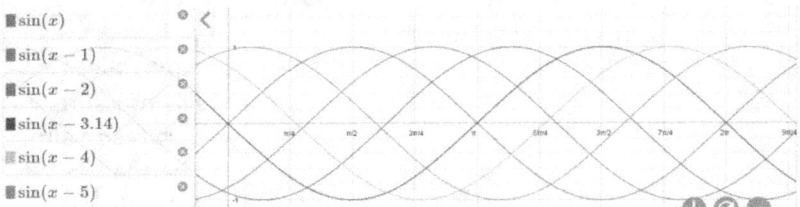

This is the situation for the gas 'state of matter', highly rotating molecules.

The Interfaces between States of Matter

The Gas-Solid Interface

More precisely, the gases, in a closed container get that extra pushback of the container wall which gets slowly spread in the speed of the gas molecules hitting others in the closed chamber.

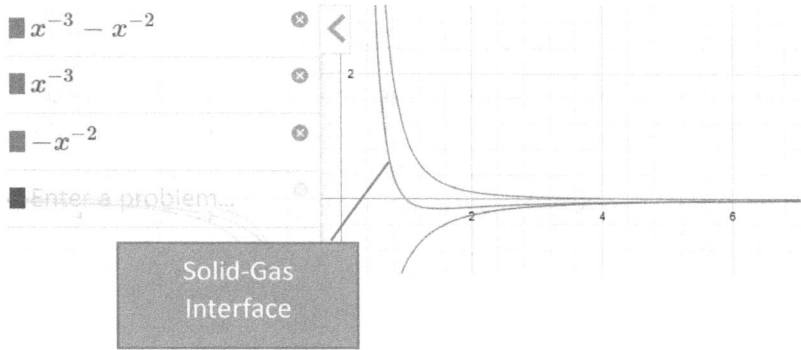

The solids are electrons out, and the gases want to expand until they reach the gravity density. That density is much less than the closed container. That is the density of outer space.

When the gas-state molecule, which is rotating enough that there are electrons-outward in every direction over the necessary time, the gas molecule comes to a solid, and hits

Solid-Gas interface	$\dfrac{kQ_1Q_2}{d^2}$

Gas-Gas interface	$\int \dfrac{kQ_1Q_2}{d^3 * \frac{8}{3}\pi(R_{ES}{}^3)} = -\dfrac{kQ_1Q_2}{d^2 * \frac{8}{3}\pi(R_{ES}{}^3)}$

So, the wall attraction, physics-negative, is greater, and that makes it rush towards the wall. Gas is $\dfrac{1}{\frac{8}{3}\pi(R_{ES}{}^3)}$ smaller.

However, that speed then brings pass that matching point, past the zero point, and into the repulsion area.

This only happens because the gases are moving and rotating so fast; only at gas temperatures, in closed chambers, do we get this repulsion process. At low temperatures, they will form liquids or solids. We see this on humid days on the window paints, and icicles on power lines in winter.

There is this zone of conflict where the stronger solid-attraction creates a whipping of the gas molecules. Forced in, then up into repulsion, and whipped back out.

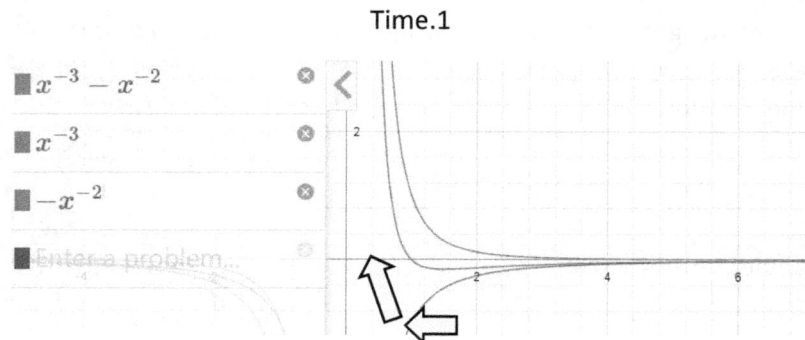

Time.1

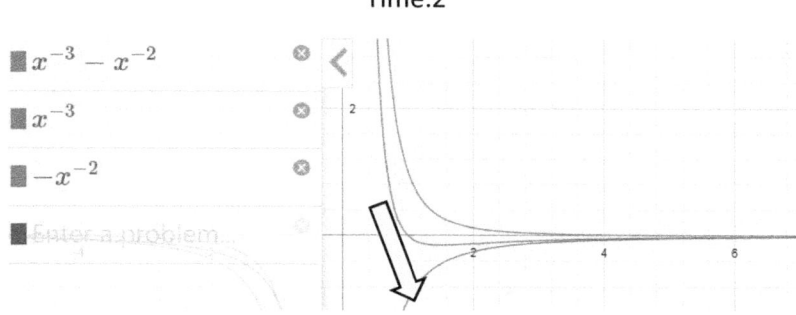

This only happens because the solid electrons-outwards are solid. The electrons have fixed settling positions, so they can act as the bouncing boards (with only minimal quantum mechanics harmonics within the range of that settling position).

This only happens because the attractive (then repulsive) force of the solid is greater than for a gas-state rotating system. It creates the constant waves crashing interface.

The Liquid-Air Interface

This same crashing-waves scenario applies at the Liquid-Air Interface.

I was experimenting with Hydrogen creation, and have an amazing experience. When we produced Hydrogen

Scrunched Cube Model of Atom

Nucleus is a Magnetic Chain-Ring

A nucleus of an atom is a structure of linked protons (black) and neutrons (white). That sequence occurs so that protons do not get too close to each other. Because if protons get near each other, this creates nuclear-reaction strength decay (and if lots of atoms do, it become nuclear explosions). Yet, for billions of atoms for billions of years existing in a stable molecule, each proton (black) is physically separated by a neutron (white), and the whole structure is continuous and stable.

For a simple molecule, these proton-neutron-proton structures can align in a chain. From that the structure has its magnetic field in the line of the magnetics links of the particles. The magnetic field go from particle to particle maintaining its strength when not separate. Most people have tried this with magnetics. When connected, they are strong.

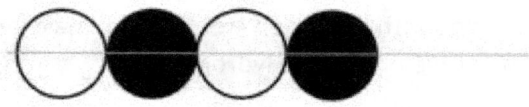

But for all larger atoms, the chain can flip and twist which might bring protons towards touching so "Bang!" proton-to-proton nuclear decay (or explosion). Alternatively, they twist into a ring, and that is actually very stable.

These can get complex with combinations of chains and rings. In the largest atoms, these extra chains flap and then just pieces get too close, then break off as radioactivity.

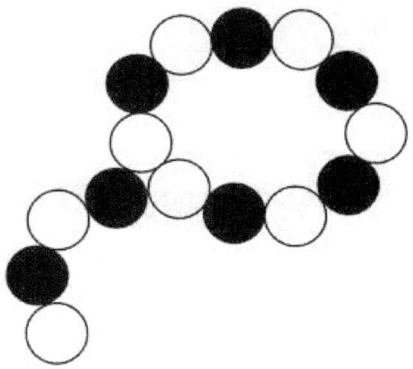

However, almost all atoms build in rings of various sizes and structures, so the complete nucleus becomes very stable with a) strong continuous magnetic chains holding them together, and b) fixed positions for neutrons to separate protons from each other. It might look like:

In the above, you even have multiple rings connected widthwise. As such, there still is neutron (green) separation of every proton (red).

Magnetics for a Ring Creates Perpendicular North-South and 'Bagel' Field Strength Shape

While the magnetic field of the individual particles may be a chained links (yellow) going around the ring, that structure actually creates another combination magnetic field that is a 'bagel' strong around the girth and weak north-south magnetic field perpendicular to the ring (red).

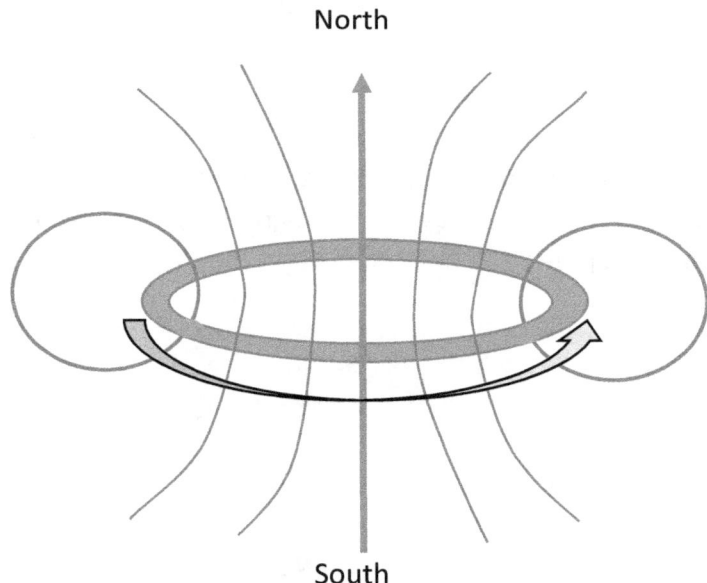

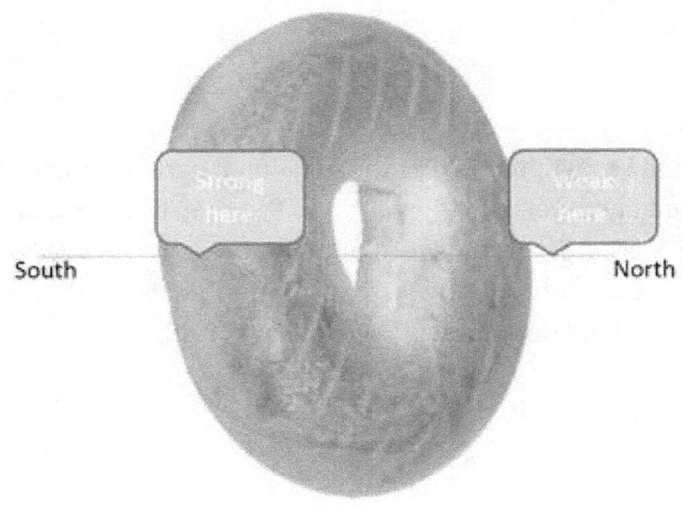

Looking at it along an axis, the strength of that field moves from strong at the bagel, at 90 degrees, towards weak at the magnetic poles.

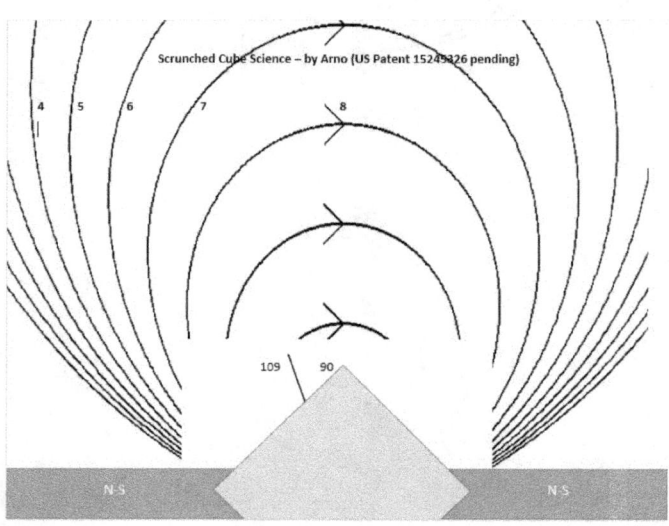

Electrons Charge Pull from Nucleus versus Magnetics Force Repulsions Creating a Balancing Point – a Shell

Everyone knows that opposite charges attract and like charges repel. That means that the electrons in the shell are attracted to the protons in the nucleus.

So, why don't the electrons just fall into the nucleus?

The answer is the other half of Newton.

Electrons are repelled by magnetics – north and south.

The most basic charge-magnetics is the balance of these two forces. Charge gives electrons the attraction to the nucleus, specifically to the proton in the nucleus, and magnetics repels those same electrons from the magnetic nucleus elements; that is, the protons and neutrons.

Therefore, you have two competing forces in Newtonian opposition. The shape is 1/distance-squared for charge, and 1-distance-cubed (in the magnetic pole direction for magnetism):

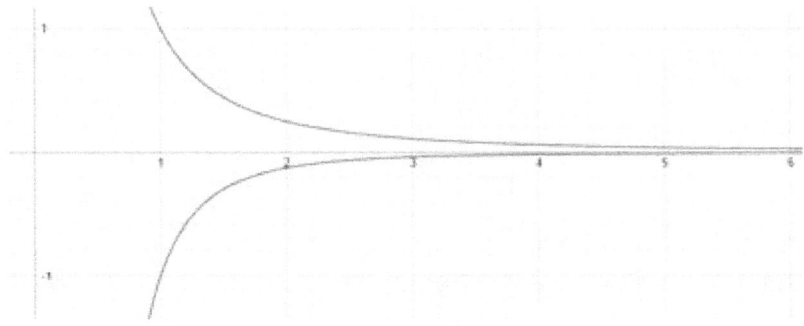

In that direction, one can look at the net force (combining charge and magnetics) to find a balancing point. The charge and magnetics offset each other the Radius of the Electron Shell.

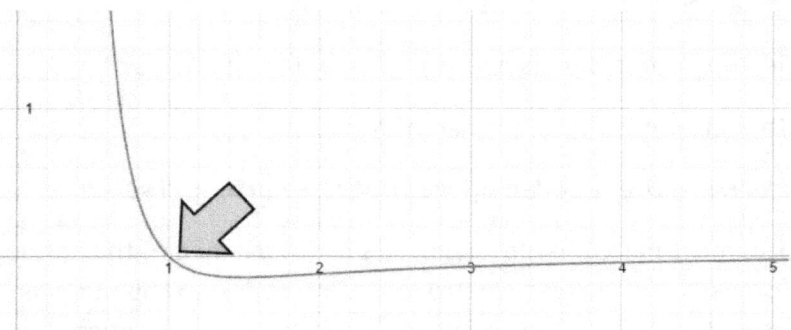

That is why electrons float in a field, a shell. They get pushed away as much as the electrons get pulled in. Wherever that happens, for each electron, there is a point of balance (yellow arrow) – the electron shells.

Adding Other Electrons Makes the Distances Different After First Two Go North-South Because Magnetic Fields Vary By the Angle

Remember that bagel. The strength of the magnetic field, the push of electrons is different based upon the angle from the end of the closest nucleus structure.

The push of the magnetic shell at the sides will be stronger, thereby the electrons will sit at that location further out.

As a result, you now have a shell structure that builds in various 3D configurations of various strengths.

Typically, these are named based upon three components:

    Shell Number   2

    SubShell (in AVSC, indicating its structure and θ)   m

    (m = the nucleomagnetics polar position)

    Number of Electrons that fit into that subshell   2

Electrons Shell – old name	Electrons Shell – new name	Direction / Structure
1s2	1m2	m = Magnetic north and south
2s2	2m2	m = Magnetic north and south
No name	2e3	e = Equatorial transition level
2p6	2c6	c = rest of the Cube
3s2	3m2	m = Magnetic north and south
3p6	3c6	c = rest of the Cube
4s2	4m2	m = Magnetic north and south
3d10		
4p6	4t6	t = Tetrahedron end cap
Different order	4f10	f = Filling five for the sphere
5s2	5m2	m = Magnetic north and south
4d10		Different order
5p6	5t6	t = Tetrahedron end cap
Different order	5u10	f = Filling five around in sphere of 1/3/5/5/3/1

Electrons Shell – old name	Electrons Shell – new name	Direction / Structure
6s2	6m2	m = Magnetic north and south
5d10		
6p6	6t6	t = Tetrahedron end cap
Different order	6u10	f = Filling five for the sphere
Different order	6v14	f = Filling five for the sphere
7s2	7m2	m = Magnetic north and south
6d10		
7p6	7t6	t = Tetrahedron end cap
Different order	7u10	f = Filling five for the sphere
6f14	7v14	v = Filling se_V_en very big sphere. _V_ery large.

At this point we have everything to Uranium covered.

Electrons Shell – new name	Cumulative Electrons
1m2	**2**
2m2	4
2e3	Transitional only
2c6	**10**
3m2	12
3c6	**18**
4m2	20
4t6	26
4u10	**36**
5m2	38
5t6	44
5u10	**54**
6m2	56
6t6	62
6u10	72
6v14	**86**
7m2	88
7t6	94
7u10	104
7v14	**118**

The old subshell naming gets replaced:

# (-)s	Prior Symbol	Prior Name	In Shell	Arno Scrunched Cube Symbol	Arno 3D Geometry Reasoning
2	s	Sharp	1-7+	M	Magnetic poles
3			1 only	E	Equatorial *[Transitional]*
6	p	Principal	2-3 only	C	Rest of Cube
6			4-7	T	Tetrahedron endcap
10	d	Diffuse	5-7	U	Big sphere
14	f	Fundamental	7+	V	Very big sphere

Each Shell, as full, then becomes a grouping of electrons in a shell. The Shells grow by Z-squared x 2 because there are two poles – north and south. Of the 3 dimensions, the magnetic is 2 – north and south; and the other two dimensions build by squares 1-squared, 2-squared, 3-squared. When full, it is rings of 1, 1+3=4, 1+3+5=9, 1+3+5+7=16 from each pole towards the equator. The spheres are not perfect though. Because magnetic fields have 'bagel' strength off the axis, then the first two (2) electrons (at the magnetic poles) tend to sit closer (get 'scrunched'), and others sit at varying distances from the nucleus based upon the magnetic strength (and the layers below). These create the bond subshells, their geometric structures, and their related angles based upon that balancing of forces:

# (-)s	Shell	Arno Symbol	Arno 3D Geometry Reasoning
	Shell 1		1/1
2		m	Magnetic poles
2	Total - Shell 1		
	Shell 2		1/3/3/1
	3	e	Equatorial [Transitional]
2		m	Magnetic poles
6	Principal	c	Rest of Cube
8	Total - Shell 2		
	Shell 3		1/3/3/1
2		m	Magnetic poles
6	Principal	c	Rest of Cube
8	Total - Shell 3		
	Shell 4		1/3/5/5/3/1

# (-)s	Shell	Arno Symbol	Arno 3D Geometry Reasoning
2		m	Magnetic poles
6		t	Tetrahedron endcap
10		u	Big sphere
18	Total - Shell 4		
	Shell 5		1/3/5/5/3/1
2		m	Magnetic poles
6		t	Tetrahedron endcap
10		u	Big sphere
18	Total - Shell 5		
	Shell 6		1/3/5/7/7/5/3/1
2		m	Magnetic poles
6		t	Tetrahedron endcap
10		u	Big sphere endcap
14		v	Very big sphere
32	Total - Shell 6		
	Shell 7		1/3/5/7/7/5/3/1
2		m	Magnetic poles
6		t	Tetrahedron endcap
10		u	Big sphere endcap
14		v	Very Big sphere
32	Total - Shell 7		

A geometric picture of each shell as they build goes:

Magnetic Poles for First Two of every Shell (1m2 naming versus 1s2)

The 1st shell is just north-south

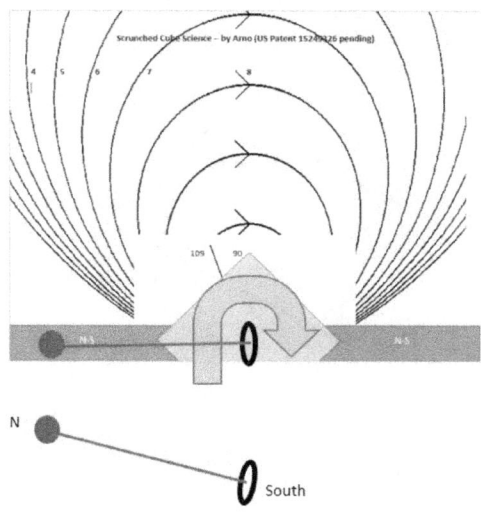

Scrunched Cube Shell 2 and Shell 3 (2c6 naming versus 2p6)

The 2nd shell is a cube scrunched at the magnetic poles. It has two (2m2) at the magnetic poles 'scrunched', and six in the rest of the cube (2c6). So, a 'scrunched cube'.

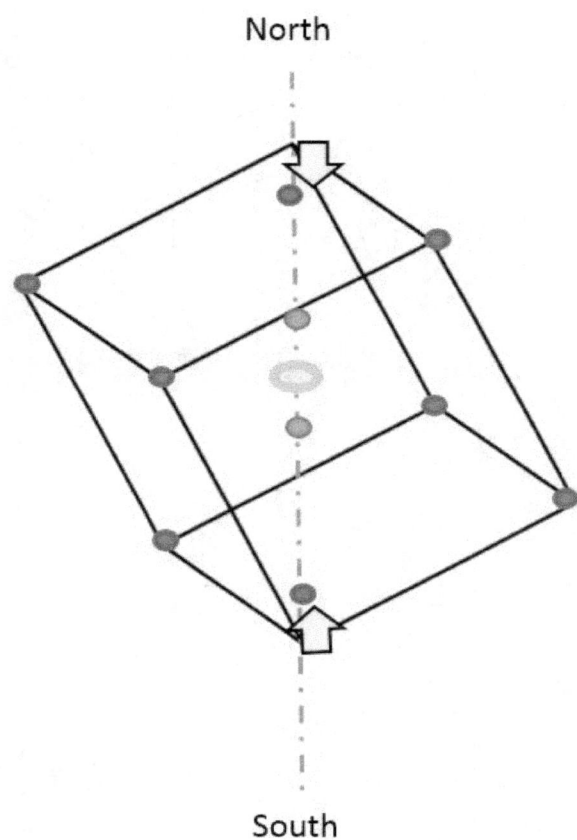

The 3rd shell is a cube scrunched at the magnetic poles. However, this electrons sit in the faces of the cube from the lower layer (with some wiggle adjustments).

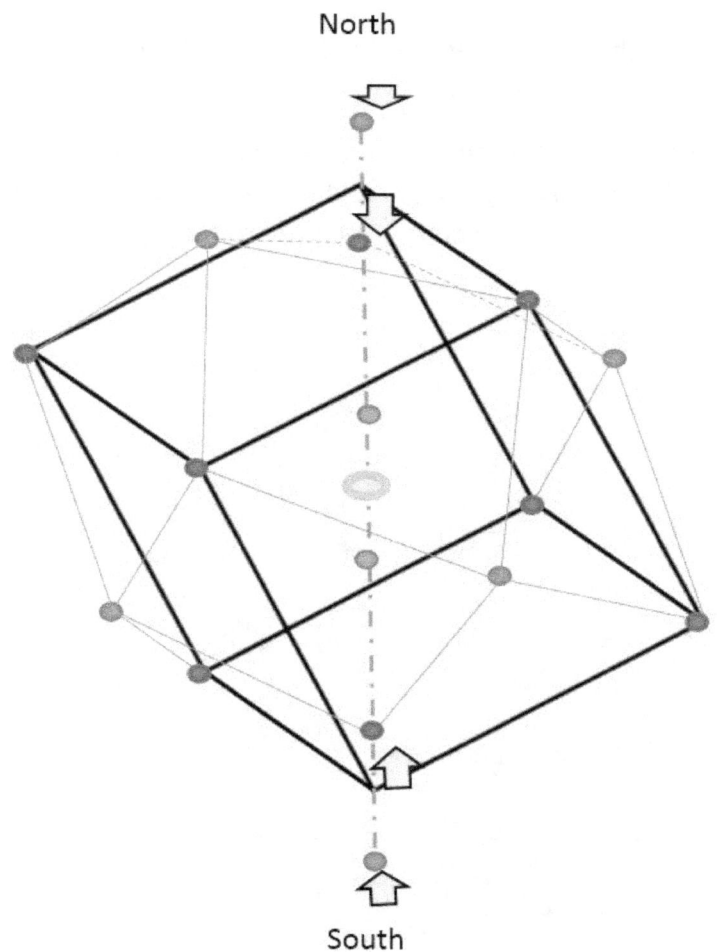

From the Magnetic Poles building (1+3+5) x 2 = 9 x 2 = 18

The 4th shell is a sphere build 1, then 3, then 5 from the each magnetic pole, and the distance scrunched towards the magnetic poles.

Therefore, the outer layer has five (5) in each hemisphere, and 2-1/2 in each quadrant.

For the below illustration, I have a electron (-) and it surrounding repulsion field (purple) so they set work from the magnetic (red) pole ½ then two more up to the equator.

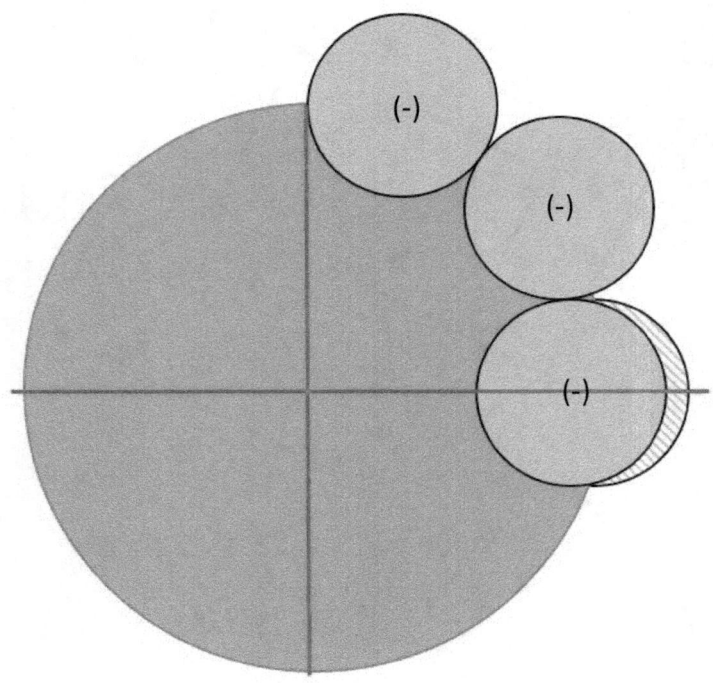

Similarly, the 5th shell is a sphere build 1, then 3, then 5 from the each magnetic pole, and the distance scrunched towards the magnetic poles

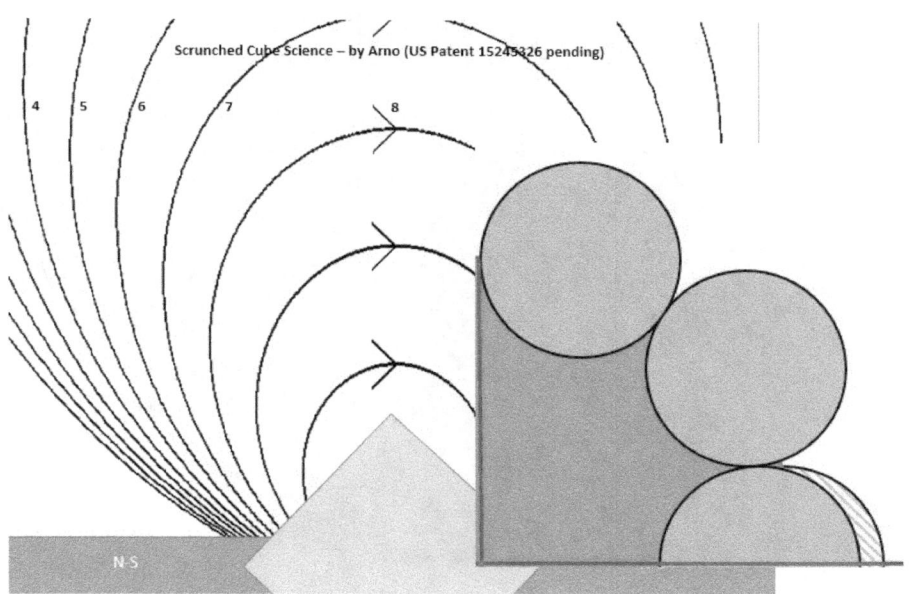

From the Magnetic Poles building (1+3+5+7) x 2 = 16 x 2 = 36

The 6th shell is a sphere build 1, then 3, then 5, then 7 from the each magnetic pole, and the distance scrunched towards the magnetic poles

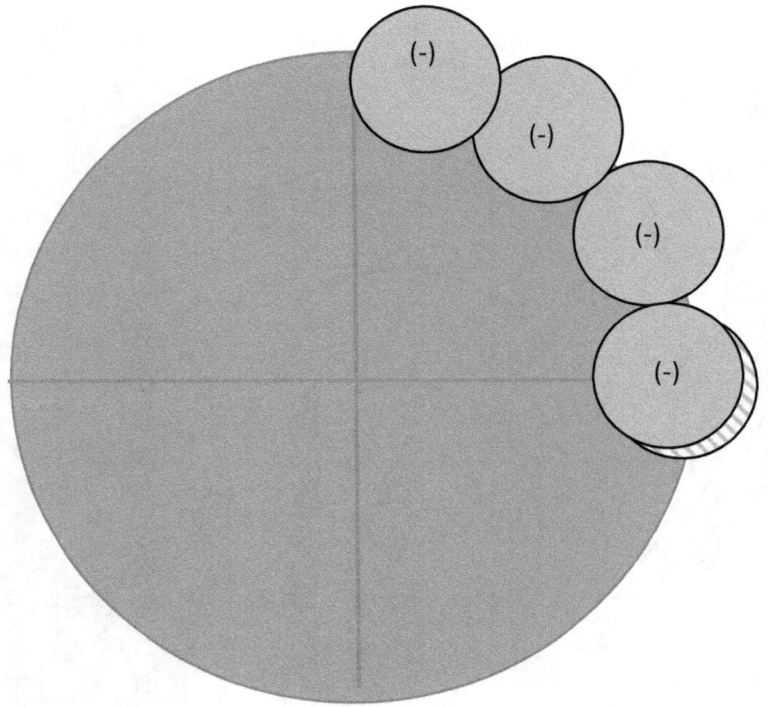

Similarly, the 7th shell is a sphere build 1, then 3, then 5, then 7 from the each magnetic pole, and the distance scrunched towards the magnetic poles. Of course, by this size, the 'scrunch' is not very different.

Shells Come In Doubles As 2nd Layer Fits in Opening of Lower Layer

From the magnetics poles, the 7th versus 6th, the 5th versus 4th, and the 3rd versus 2nd shells offset by ½ so an extra layer can fit between before building to a larger set of electrons.

For Shell 2 and Shell-3, the build is easy to see, the 1 is at the equator, and the 3 electrons section is closer to the equator:

From the Magnetic Poles building 1 x 2 = 2

From the Magnetic Poles building (1+3) x 2 = 4 x 2 = 8

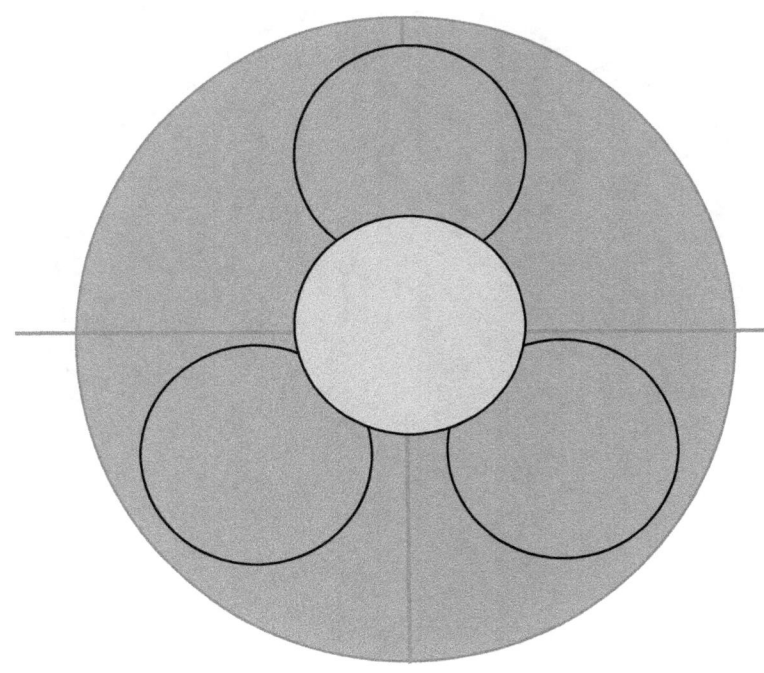

From the Magnetic Poles building (1+3+5) x 2 = 9 x 2 = 18

It gets more complex from 1/3/5.

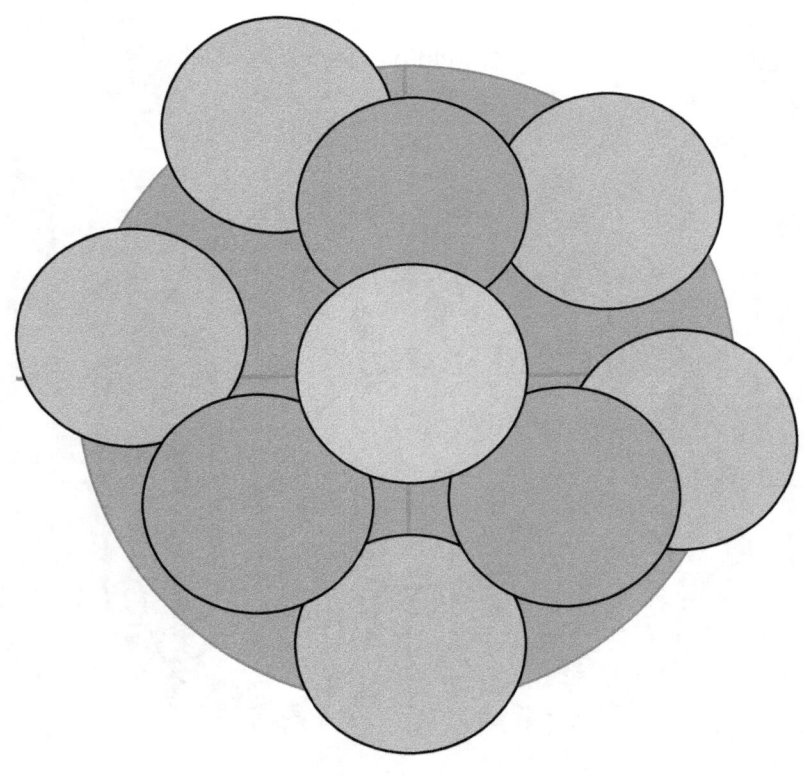

From the Magnetic Poles building (1+3+5+7) x 2 = 16 x 2 = 16

It gets more complex from 1/3/5/7.

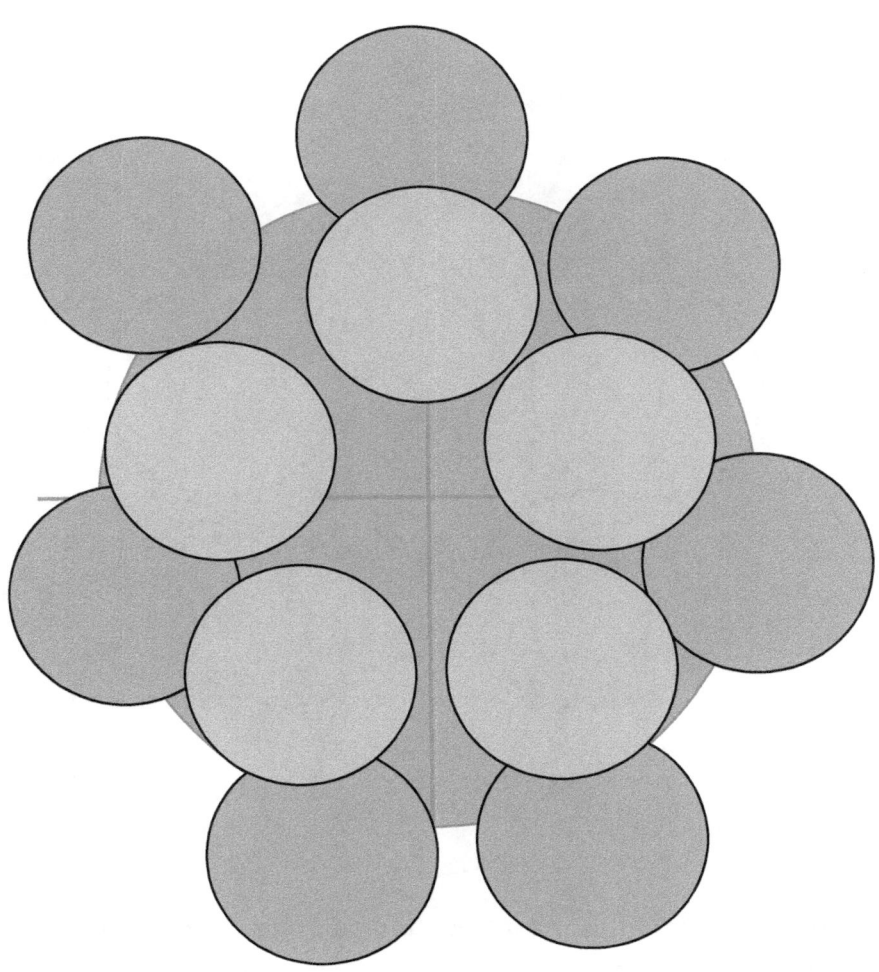

After a 2nd layer, there is no open spots (by balancing ½ rule another balancing goes back to the first which is already full) so the next layer become a new configuration (3 > 5, 5 > 7). Plus, there is plenty of extra room. Therefore, a new layer – at a slightly different spherical longitude with more electrons (and their electron shell repulsions) occurs.

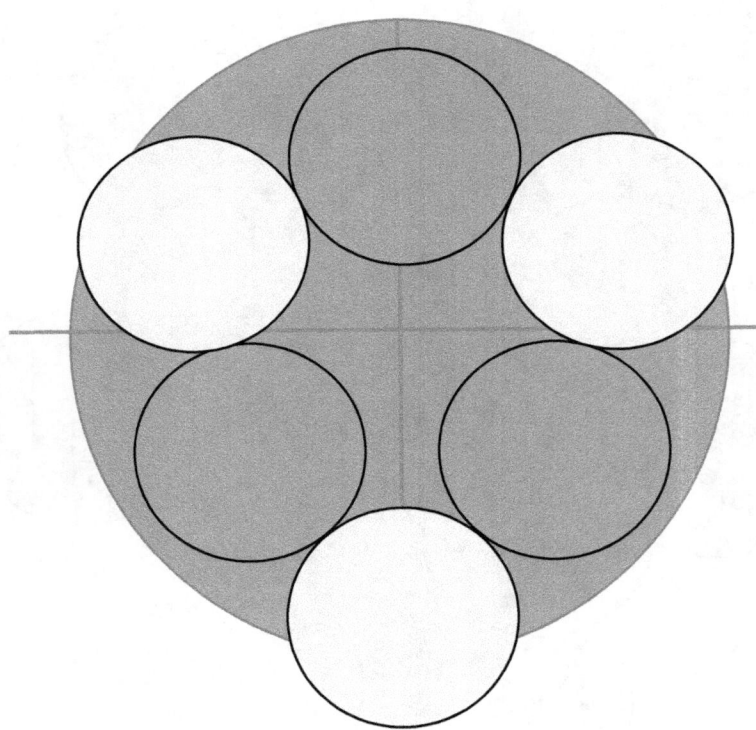

You can see that the next layer, the odd-numbered Shells, 'fits between the layer' below (and effectively reverse longitudes in the opposing hemisphere). Of course, two reasons make the next, next layer not fit between the even layers.

1) Fitting between an odd layer; that brings you right back to the base (even) layer, so repulsion from below exists at that exact position. That give the chance for slightly different pattern to actually 'fit' better.

2) There is more room so you can change in size (the 'squared-rule'). The same spacing of stuff grows by the square. You can fit two extra – it's just the math of any square.

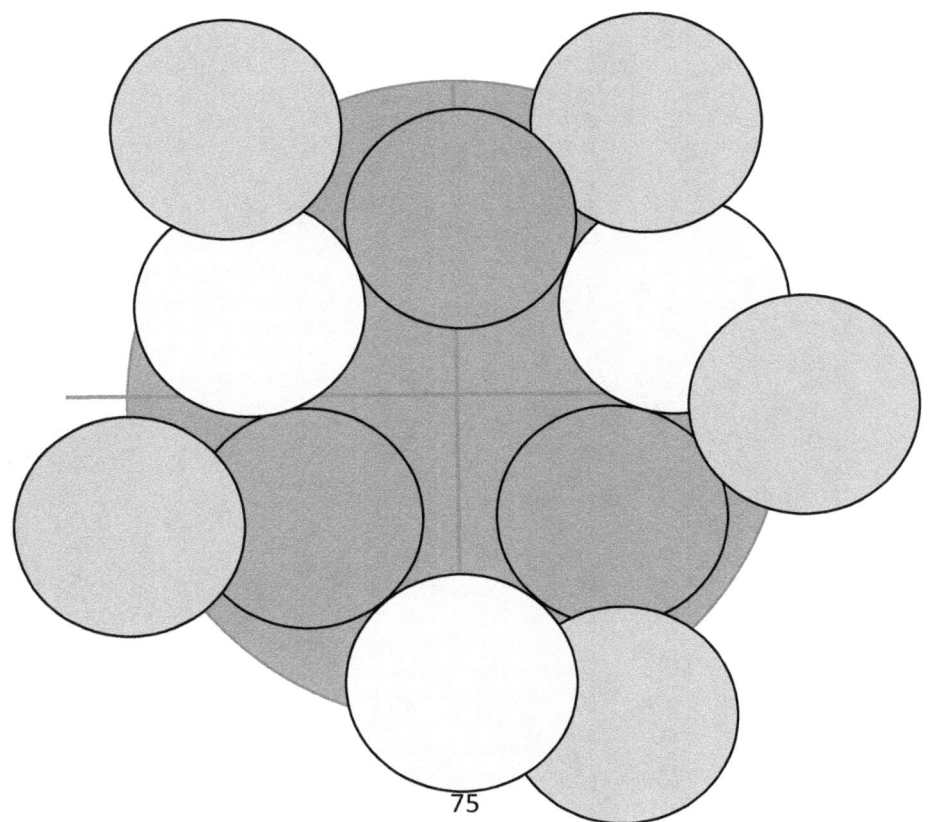

Adding Other Electrons Repulsion to Charge-Magnetics Balancing

Double Shells Creating Room 2 poles x N squared

Transitions Metals –Transitional EndCap Pyramid (4m2+4t6 vs 4d10)

Replacing Pauli/Aufbau Logic with 3D Geometry Transition Models

Arno Chart of Elements

The current periodic table uses a different building structure. The Arno Scrunched Cube Atomic Model would group elements differently.

The current model orients more toward right filling, but the Arno Scrunched Cube model orients more toward both ends. Therefore, both ends have common columns, but the middle have better grouping. Therefore, the Lanthanides and Actinides are broken up and re-built with the left focus elements.

1) The Noble Gases would not change as those are full shells in all models.

2) The Halogens that are near-full also are matching in the two systems.

3) The 2nd shell is treated entirely differently and separately as the properties are not columnwise.

4) The Alkali Metals and Alkaline Earth Metals from Shell 3 up would remain.

5) However, the 2nd Shell first three elements are more similar and should get grouped widthwise, not columnwise.

6) The transition metals would would end at t6, not the traditional d10. The change of

7) The first six Lanthanides and Actinides would fall in the EndCap group.

Therefore, the Chart of Elements would group like:

Full and Near Full Outer Shell:	
Nobel Gases	Build from the right
Halogens	Build from the right
Just Building Outer Shell:	
Alkali Metals	Build from the left
Alkali Earth Metals	Build from the left
Endcap Magnetic Metals	Build from the left
High-Melt Metals	Build from the left
Non-Multiple Shell As Oriented:	
Hydrogen	Small Molecules
Equatorial-Reactive	Small Molecules
Scrunched-Cube-Reactive	Small Molecules
Mid-Range Structures:	
Electrically-Most-Active Metals	Longitudinal Electrons
Poor Metals	Semi- Malleable
Metalloids	Malleable

1	1	6 or 16	grows		varies	varies	1	1
001-H								02-He
Equatorial-Reactive					Scrunched-Cube-Reactive		009-Fl	10-Ne
11-Na	12-Mg		1 x		1 x	2 x	019-Cl	18-Ar
19-K	20-Ca	6x	3 x		2 x	2 x	35-Br	36-Kr
37-Rb	38-Sr	6x	4 x		2 x		54-I	54-Xe
55-Cs	56-Ba	16 x	5 x		1x		85-At	86-Rn
87-Fr	88-Ra							
1-Polar Alkali Metals	Both Polar Alkali Earth Metals	Endcap High Melt Metals	Electric Metals		Metalloid	Non-metal	Halogen	

As such, the traditional d10 (or 14f subshell) falls into different buckets. The activities of elements in the new blocks act quite differently.

			6 x	2 x	2 x			
			6 x	2 x	2 x			
			16 x	2 x	2 x			
			16 x	2 x				

1) The electromagnetic spectrum changes dramatically after then 6x section, not 10x

2) The melting temperature peaks for the High-Melt Metals, and it much different in the other categories

3) The magnetic profile changes dramatically after the 6x section

The basic reasons for this is the geometric placement of the electrons in the proposed t6 shell, then the u10 shell.

The t6 shell sits at the 2nd subshells close to magnetic poles (m) as endcaps m2 + t6. This makes the atomic structure oblong, not spherical.

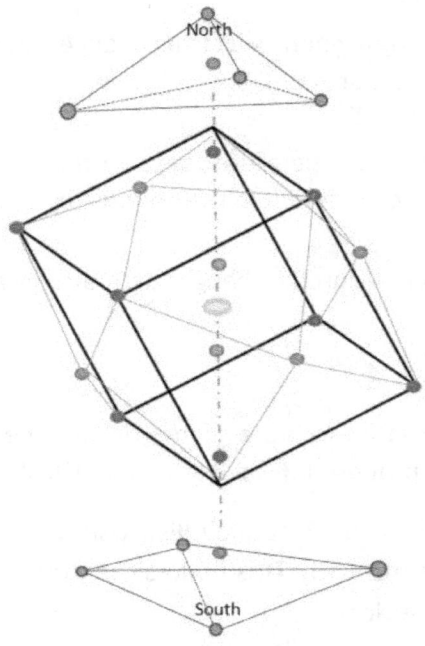

This configuration is amazing. You have the lower shell exposed almost fully, and the next shell is only at the magnetic poles. For ionizations, you have dozens of similar endpoints where electrons can move from one to another creates almost 100 different reasonably equally available release distances (bandwidth) for those subshell exchanges.

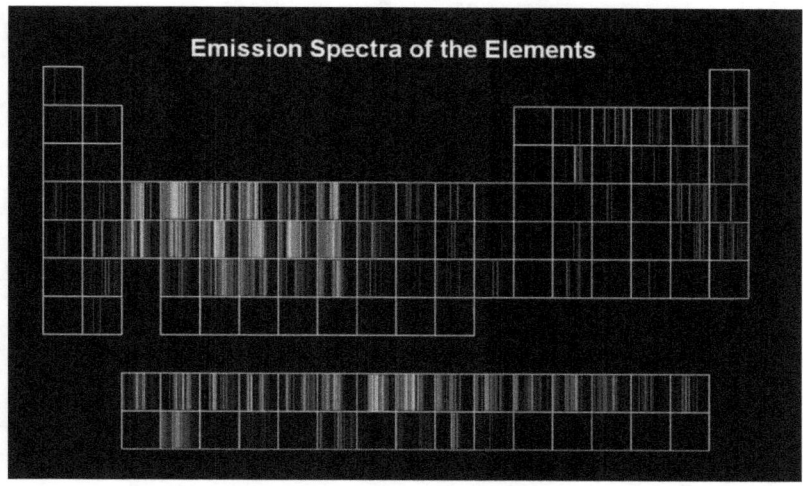

Notice how the variety of spectrum stops at 8 (m2 + t6) consistently on each row of the current periodic chart.

Note that at the Lanthanides series, the endcap is not 1+3, but 1+3+5 such that the endcap is similar, but a layer bigger. Remember that the underlying layer at Shell 6 and higher is already 1+3+5 so that extra layer still does not make the endcap any bigger than the bulk of the base set of electrons shells. It is still an endcap.

However, once you get past that endcap, the extra electrons truly sit exposed. They sit at longitudes towards the equator, and further have 180 degrees of direction where incoming particles or energies might move their position. At a huge break, the relative of a dozen exposed electrons becomes just the one.

Now, in the best scenario, the most electricity, 029-Cu Copper, there are three (3) electrons out at this transitional equator location. The 026-Fe Iron has zero.

The ones 27-Co Cobalt to 28-Ni Nickel get more electrically active up to 029-Cu. 27-Co Cobalt is active with one electron at the equator; 28-Ni Nic kel is more active with two electrons at the equator, and 29-Cu is the most electrically active element of all.

Past 29-Cu Copper, the angle (at 4 at the equator) becomes 90 degrees, but the equatorial electrons get pushed into u-subshell location where they can stay three around at 35 (not 90) degrees at one pole and one at the other pole – three around the latitude more towards the poles. Equatorial tradition subshells can only handle three (3) electrons. Above three (3), then the fourth electron at 90 degrees pushes electrons out such that a non-equator location is better.

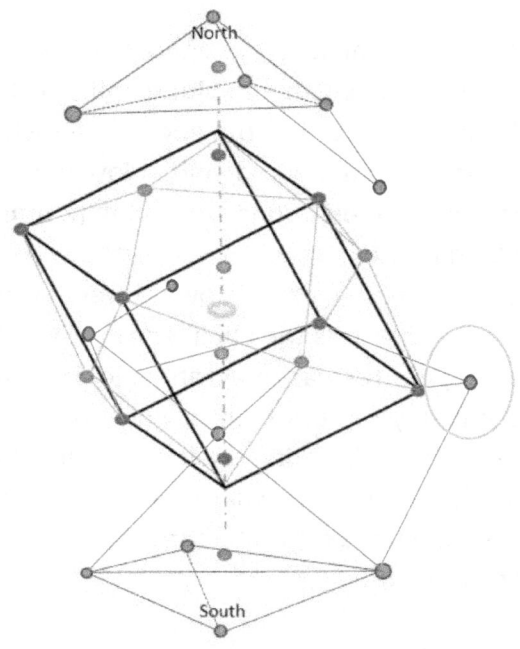

This new position:

- Takes up almost all the opportunities for electromagnetism and blocks multiple lower frequencies. Unlike the endcaps, which are open positions to the world, and at lower energy levels, the equator electrons are in high (2 time from 1+ cos-90-squared) magnetics and further take all the open energy of many, many lower layers. Only once the 4u shell starts is the Shell-3 not active. Those are still there, but the vast majority of ionization occurs at the highly-repulsion-energy electrons which also are now above, blocking the inner shell more fully. The vast majority goes where the energy is quite low, and the directions for that energy are over 180 degrees of the horizon.

   For endcap configurations, each new electron has about 20-30 degrees of exposure. It has lower level c6, u10, as well as outer shell m2, t6 electrons all with about the same likelihood of creating spectrum emissions. The lower shell, and the outer endcap all have exterior exposure. Therefore, all the varities of electrons exchanges express in generally equal electromagnetic spectrum of those multiple distances.

- Increases the last electron repulsions such that it can move much easier in electricity interactions. They are open to move to the next electrons at over 180 degrees.

Electromagnetics and the Scrunched Cube

Charge-Magnetics Force and Distance

Distance and Electron Shells – One Atom

Volume Generally Linear to (Z+N) Magnetic Force of Protons and Neutrons

Overlap

Distance and Molecular Bonding

Angles and Electrons Shells – One Atom

Angles and Molecular Bonding

Larger Atoms No Angle, More Underlying Layer Repulsions

Movement of Molecules

Gravitational Orbit versus an Electron-Shell Orbit

Rotation and Molecules

Non-Bonding Cohesion or Adhesion

Specific Chemical Processes

H2O Water

CO2 Carbon Dioxide

Water has two double bonds set opposite. However, those double bonds are perpendicular to each other, and further those bonds are not in the direction of the magnetic poles for any of the three molecules.

A CO2 molecule is a carbon (black) with two oxygens at 180 degrees, but with double bonds sitting at 90-degree cross orientation (blue section changed to match the 'scrunched-cube model versus showing bonding in line as per the original textbook).

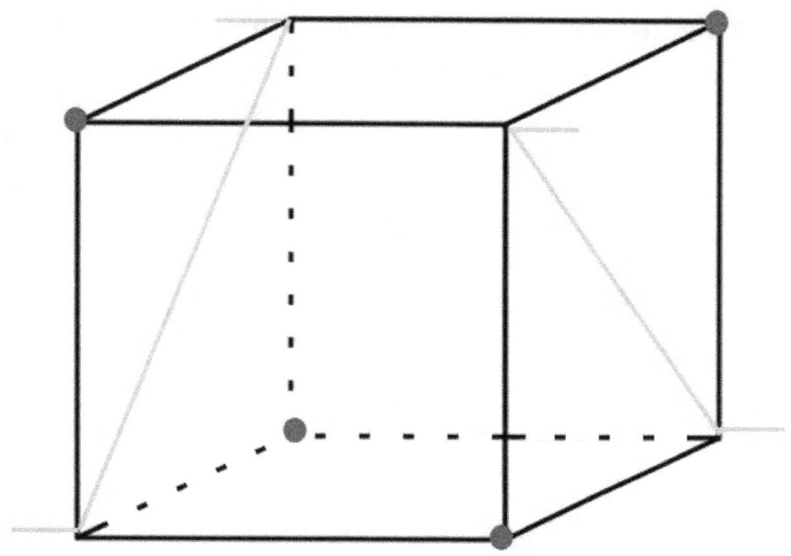

However, this straight out is not the natural position, that makes the CO2 bonds stressed.

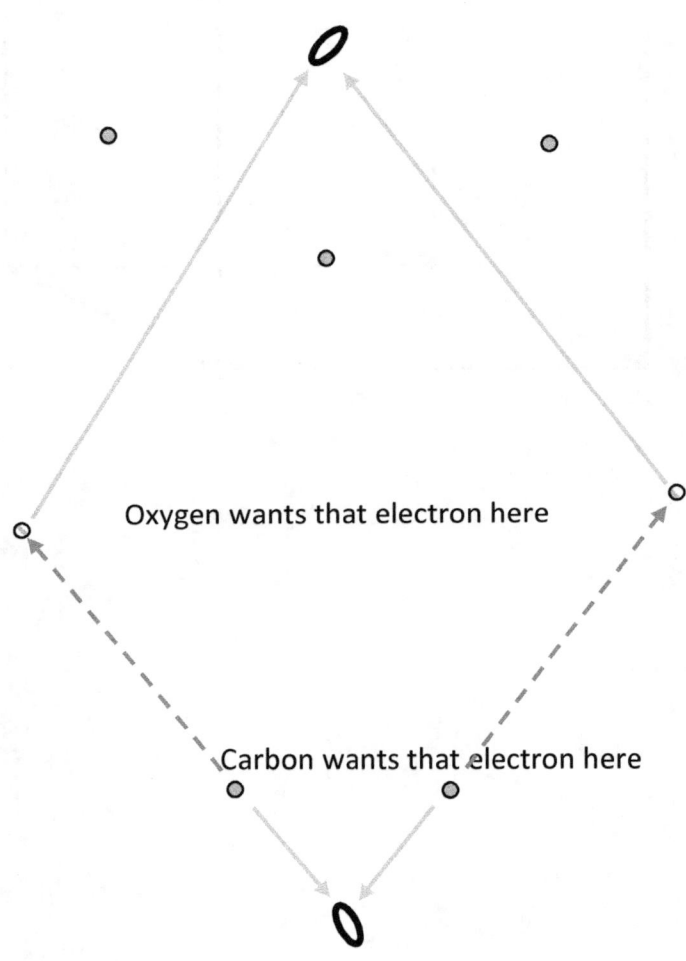

Therefore, there is the extra pull (red) on the electron in this double bond.

That stress is the angle of the double bonds versus the tetrahedron angle.

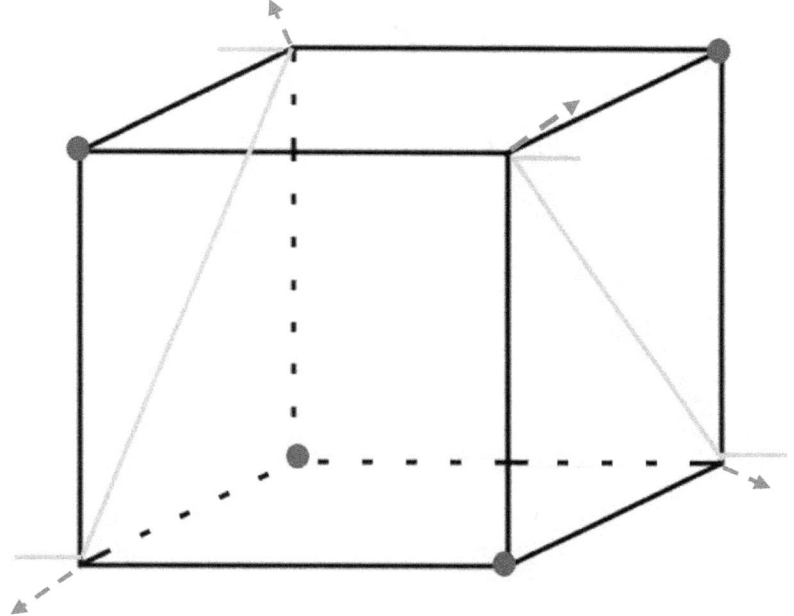

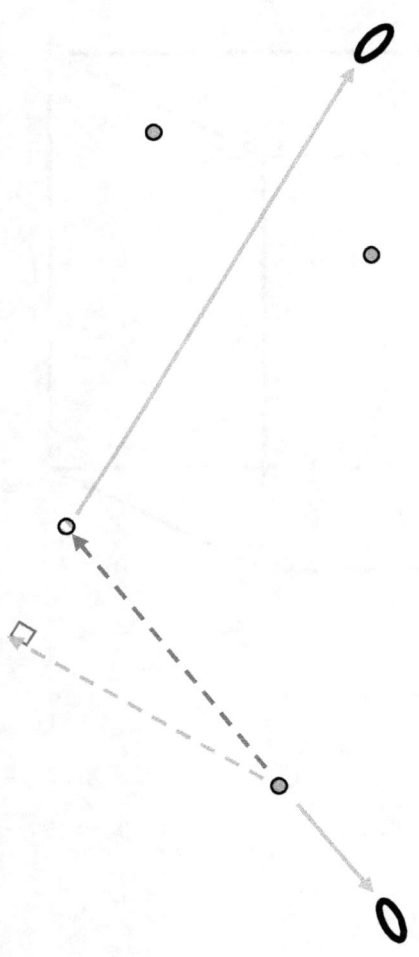

Further, the magnetic fields of the three individual atoms are not aligned with the bonding orientation (as is usually the case).

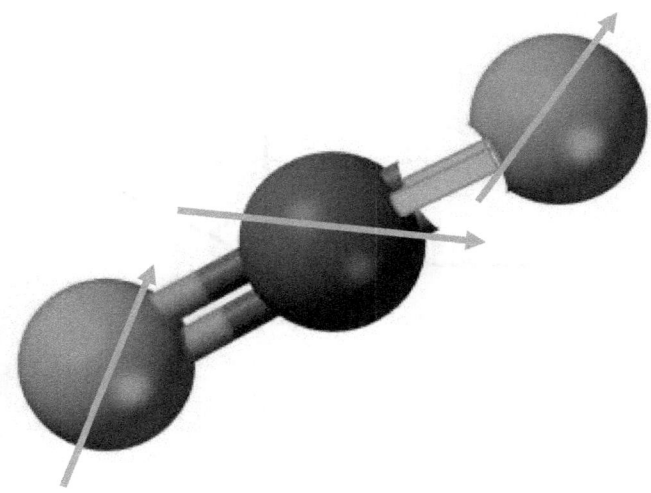

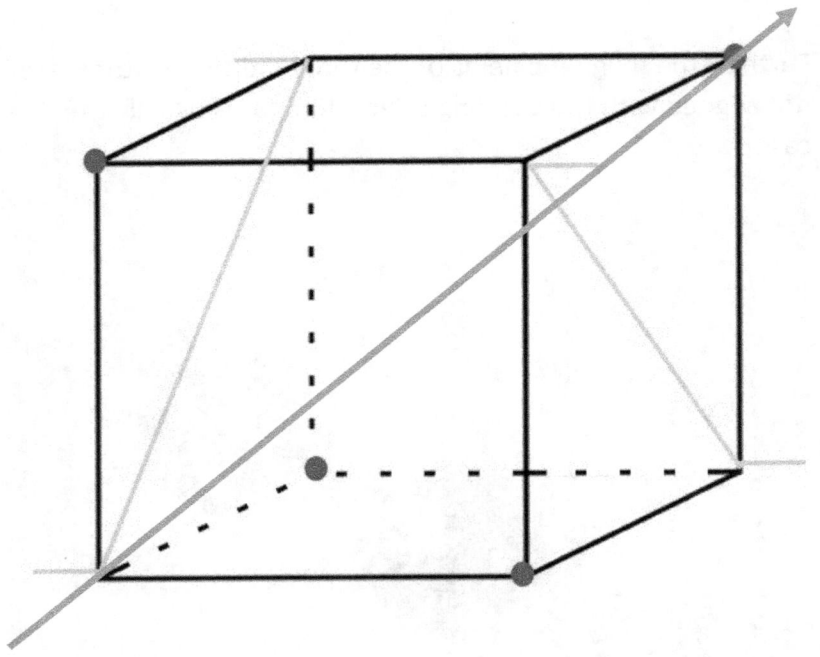

For a CO2 molecule, you have a Carbon in the center. That carbon has four bonding positions at the corner diagonal of a cube (which is a tetrahedron). The entire structure is 'scrunched' slightly toward the magnetic pole end while staying in the 4-diagonals-of-a cube structure.

The below view is from the equator of the nucleus magnetic field.

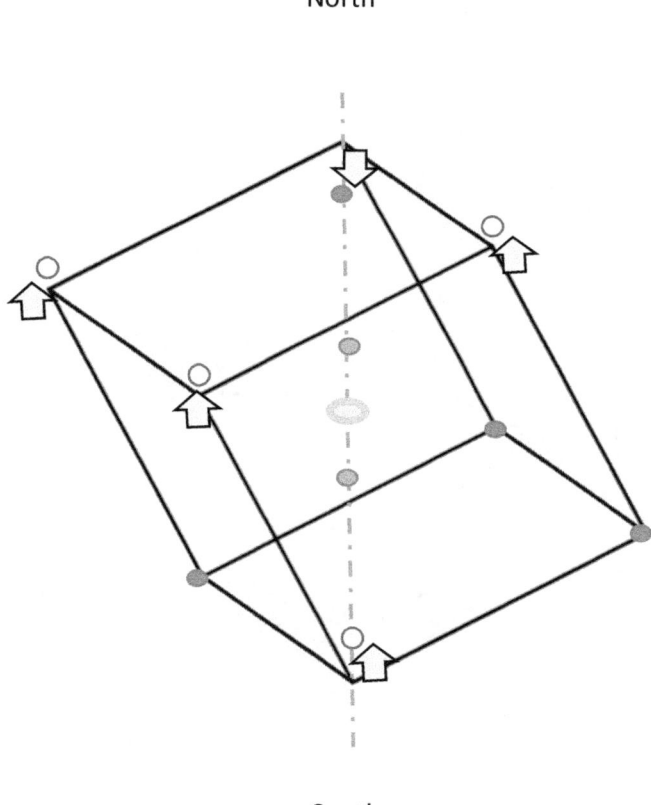

However, if looked at from the direction of one of the bonding locations, the structure looks very balanced. This a view 'lookup up' if you are having trouble. The bottom bonding position is 'down' from this view. The tetrahedron is in dashed lines a pyramid base to the 'bottom' even further.

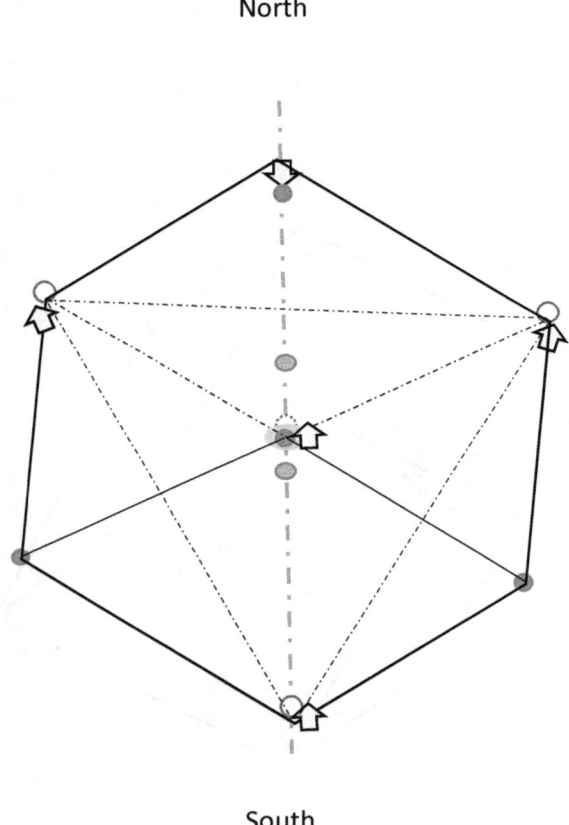

North

South

In this view, the magnetics are running directly toward the viewer. It runs from the top corner (red hashed), through the nucleus (in the middle), to the bonding position (white) at the bottom.

Notice also that the bonding positions actually move slightly with the 'scrunch' as the repulsion along that magnetic line is less, so the bonding atom moves by XX, and the other bonded atoms move by that same distance as at 3.xx times that distance, the balancing of repulsions makes them balanced for a carbon atom.

The 008-O Oxygen atom has different scrunching as the outer shell has six location filled. Since only two bonding positions, they get 'scrunched' up and out so the angle is not 109.5, but 104.5.

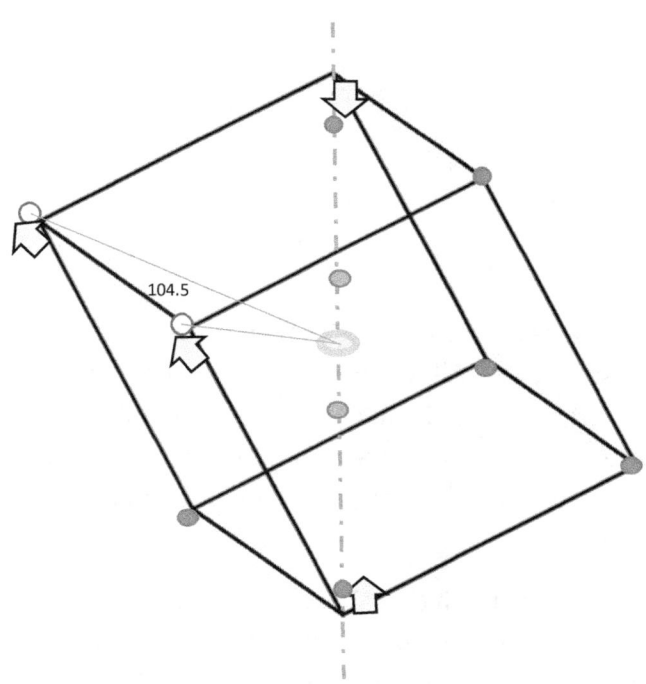

This would seem to work. Both are double bonds across the face. They generally line up.

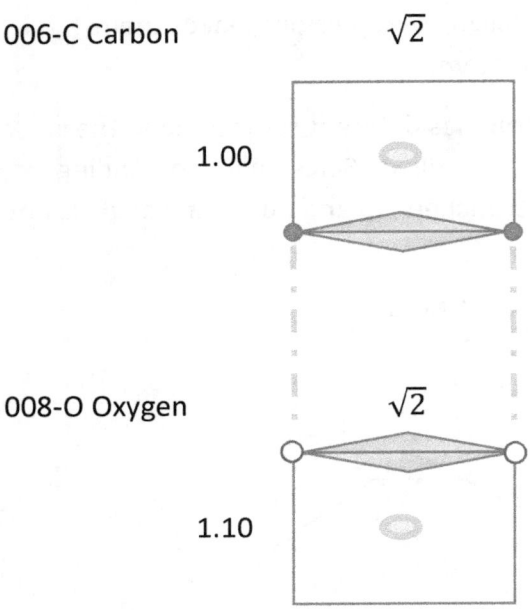

006-C Carbon

$\sqrt{2}$

1.00

008-O Oxygen

$\sqrt{2}$

1.10

However, the rest of the atoms makes these sit at 3.xx x molecule size apart. All these bond angles would work great, except both the C and O are relatively full. That means four O's around a C has too much repulsion from the five (5) electron shells. Further, what would happen to the extra O 2nd bond.

As a result, C likes to give up 2 towards each O. This is a 006-C Carbon in the middle, but as a contributing double. The basic view of the bond is a double across the diagonals both for the 006-C Carbon, and for the 008-O Oxygen.

However, you can see tension if viewed by the two atoms. The receiving 008-O Oxygen wants the electrons to sit at 104.5 degrees apart at 3.xx times the atom radius distance (Bohr's radius x (nucleus magnetics)^(2/3)). However, the 006-C Carbon has a natural distance of 107.5 degrees at the atom radius:

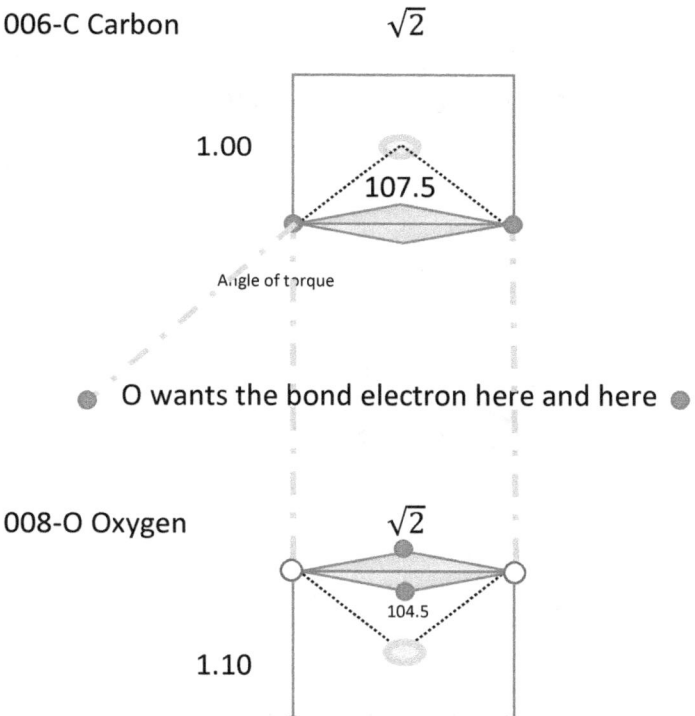

This creates a natural tension that can get attacked in the present invention by pressure to:

- Move the 008-O Oxygen in the direction to make the electrons replace the bonding positions which drives the bond into breaking. This an angle specific to the magnetic orientation of the O.

- Move the 008-O Oxygen away which increases the torque until it drives the bond into breaking.

- Insert an 001-H at the correct angle

Note that the angle of magnetics for the 006-C Carbon in the center and the 008-O Oxygen atoms are different.

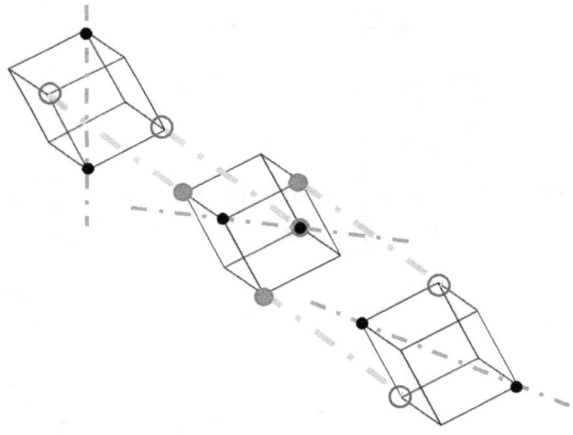

You have no chance that the magnetics of the 008-O Oxygen at each end are the same. The 008-O Oxygen molecules has 4 pairs of potential magnetic orientations. All four pairs have a location on the face showing toward the 006-C Carbon. The two bonding pairs (A and D) with the polar opposites are not orientation possibilities.

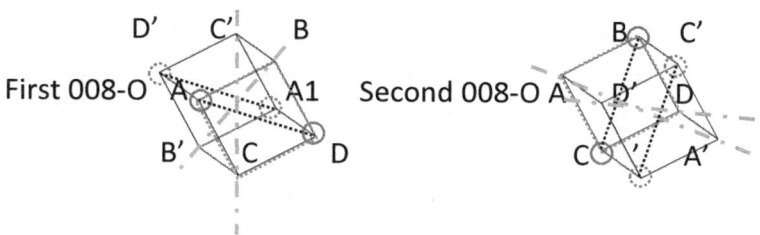

Because the contributing electrons are at diagonals at opposite corners in the cube > tetrahedron structure. The 2nd 008-O Oxygen it twisted by 90 degrees in the bond, even if the bond direction are 180 degrees from each other.

The magnetic base of the outside 008-O Oxygen are also twisted at 90 degrees. They can NEVER align. Both Oxygens have the magnetic angle NOT through the bonding positions, and the magnetics must go through two opposites. Therefore, the bonding angle must between the two remaining.

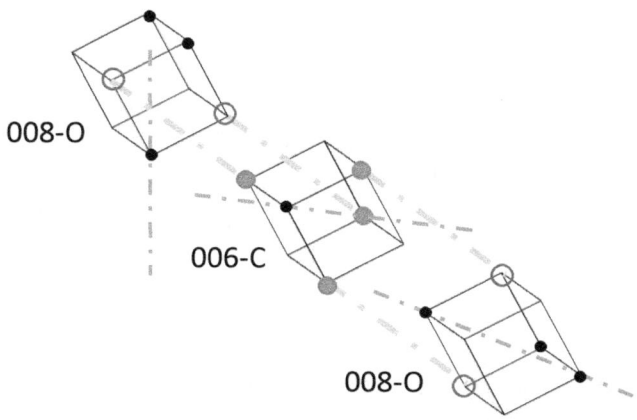

That means the magnetic angle of the two end 008-O Oxygen are at 90 degrees to each other, and both those are at 114 degrees 114 degrees to the direction of the total magnetics, but at 90 degrees to each other. Therefore, a limited action in that direction breaks the bond more easily that overall pressure. Pressure to make one align like the other will pressure the bonds already under torque stress.

Yet, if you look at the magnetic field from the top of the three, then you get closely related orientations.

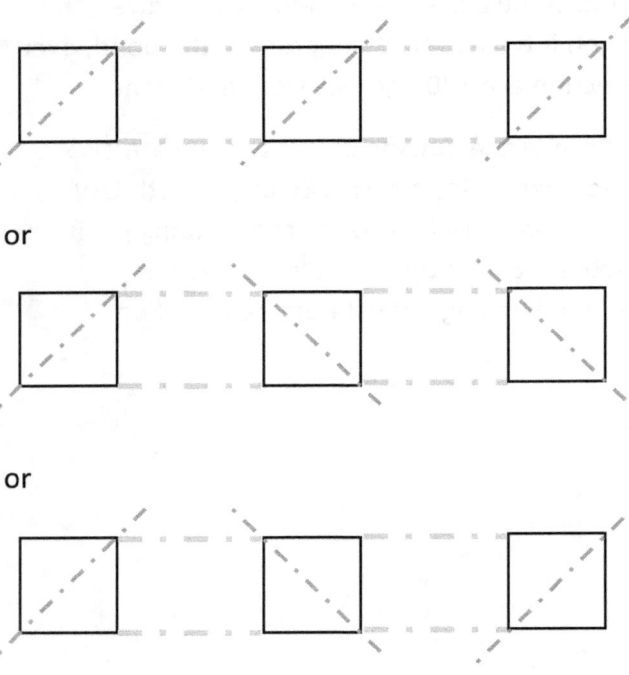

So, at 45 degrees from the strongest magnetic orientation, the three atoms in the CO2 molecule align well.

1) all 3 align magnetics, or

2) two (2) of the three (3) atoms will align magnetics, or

3) the two ends align and the 004-C Carbon in the middle is at 90 degrees versus the outside.

In either case, the potentials for applying limited force for a 45 degrees base structure.

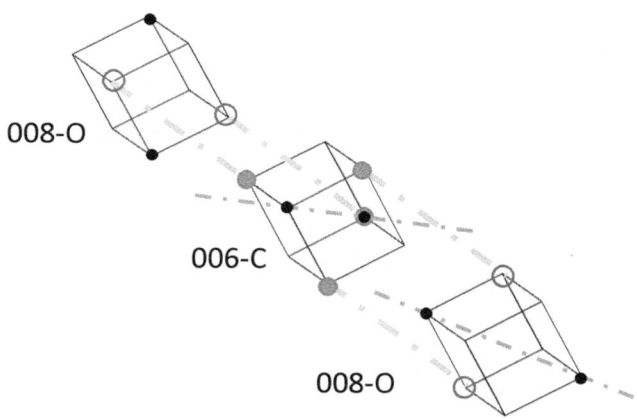

008-O

006-C

008-O

However, if looked at from the direction of one of the 006-C Carbon bonding locations, the structure looks very balanced. If you look from one corner through the nucleus to the opposite corner, those align. From that direction, the other six (6) cube positions are perfectly 60 degrees separate, although three (3) are front and three (3) are back in the cube. Further, the mangentic orientation is at exactly 90 degrees.

North

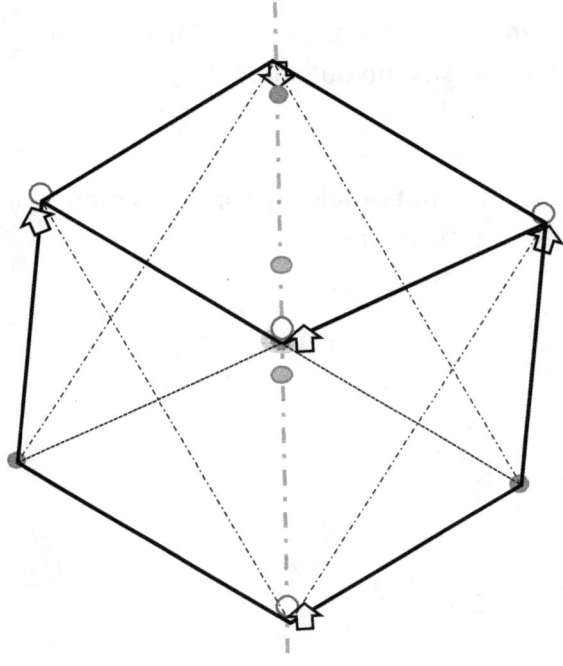

South

026-Fe Iron (and 008-O Oxygen)

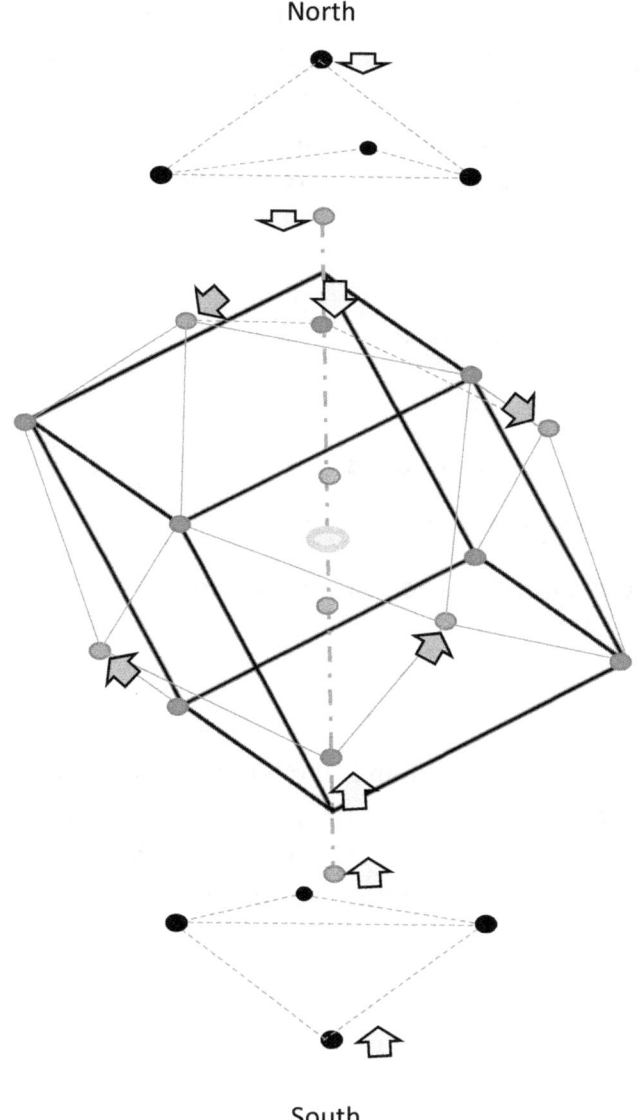

There are not too many places for bonding for 026-Fe Iron. The electrons are evenly distributed. The new electrons form an endcap, and those electrons are near the magnetic poles so they have less repulsion (so are more tightly binding). As a result, 026-Fe Iron does not bond very much or easily. As much as one electron might get pulled to another molecules, the surrounding electrons would then align to get repulsed by the electrons around the bonding position.

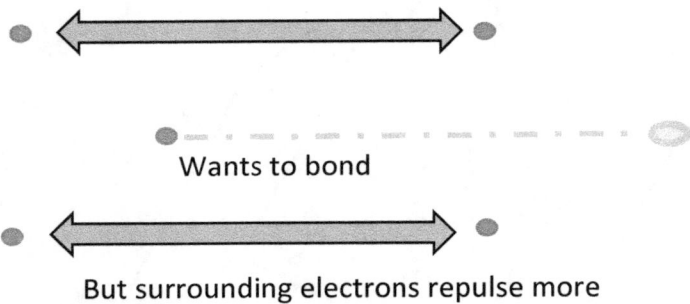

Wants to bond

But surrounding electrons repulse more

The most notable exception is the 008-O Oxygen that wants a double bond. In that case, the polar electron and one face electron are both sticking out. Even better, they are sticking out about the distance of the double bond.

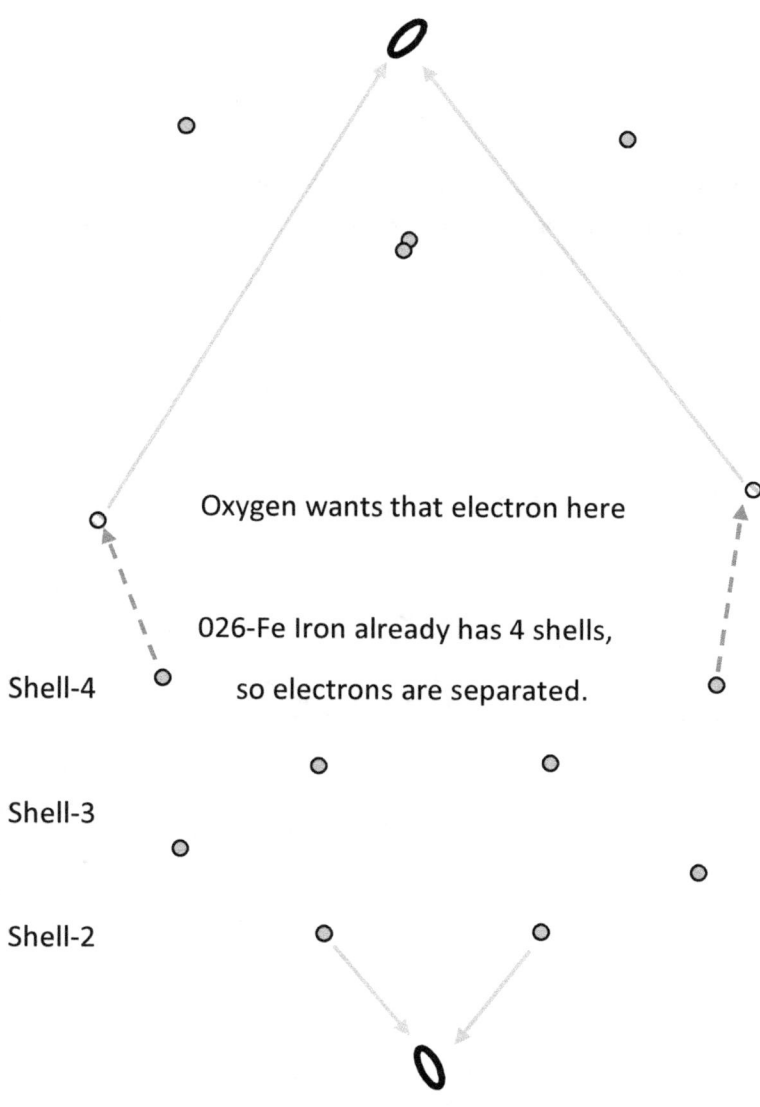

The torque on this bond is not very much. The outside electrons act as the double bond. The best choice is the pole and cube-fact.

H2O – Water

Water is Scrunched Cube 008-O Oxygen with two 001-H Hydrogen at 114 degrees from magnetic poles, and 104.5 degrees from each other.

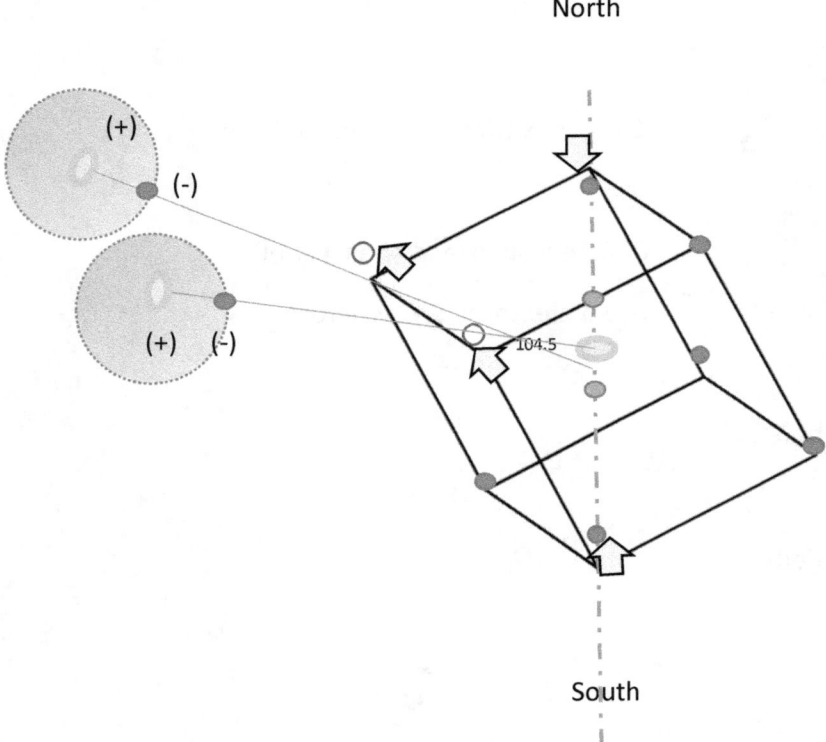

Yet, the electrons rarely rotates about the hydrogen nucleus proton. That would break the bond. When the proton is closer, the system has net repulsion.

Instead, the system rotates with the Oxygen which is a different sort of 'orbit'. As the system turns, the electron would rotate in this manner relative to a distance observer, but not relative to the Oxygen atom. To the Oxygen, the electron is generally closer.

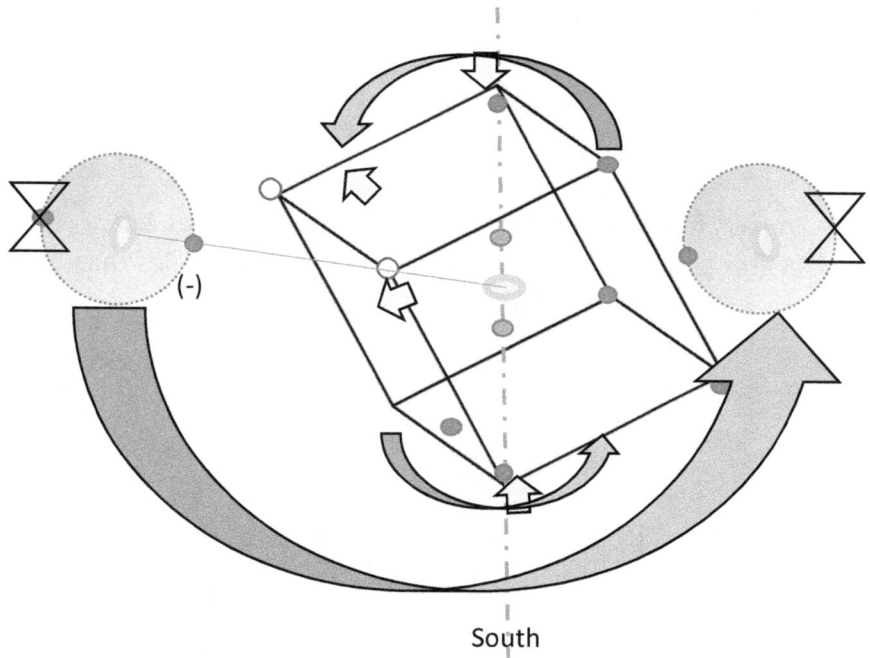

South

Some secondary 'quantum-paths' based upon a 'top-like' spin on the axis between the 001-H Hydrogen proton and the 008-O Oxygen nucleus can occur. Remember that all particles are free in 3D, so a combination of forces creates a set of 'quantum-paths' that are viable. There still is a statistical result where electrons are not 100% in the exact 104.5 degree path to the hydrogen.

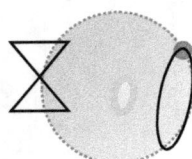

In fact, this wobble is common as water dissolves salts. Water sits in a class of Exterior hydrogen molecules that have this wedge sticking out of a proton.

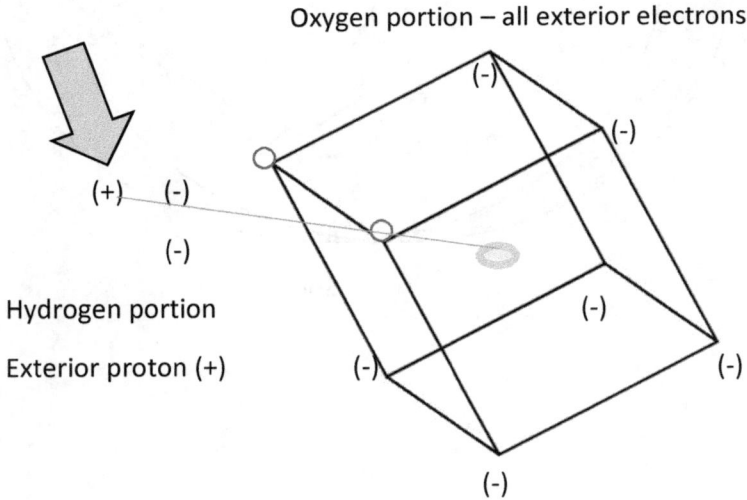

Whereas, an O2 molecule of two 008-O Oxygen atoms on the outside has only electrons.

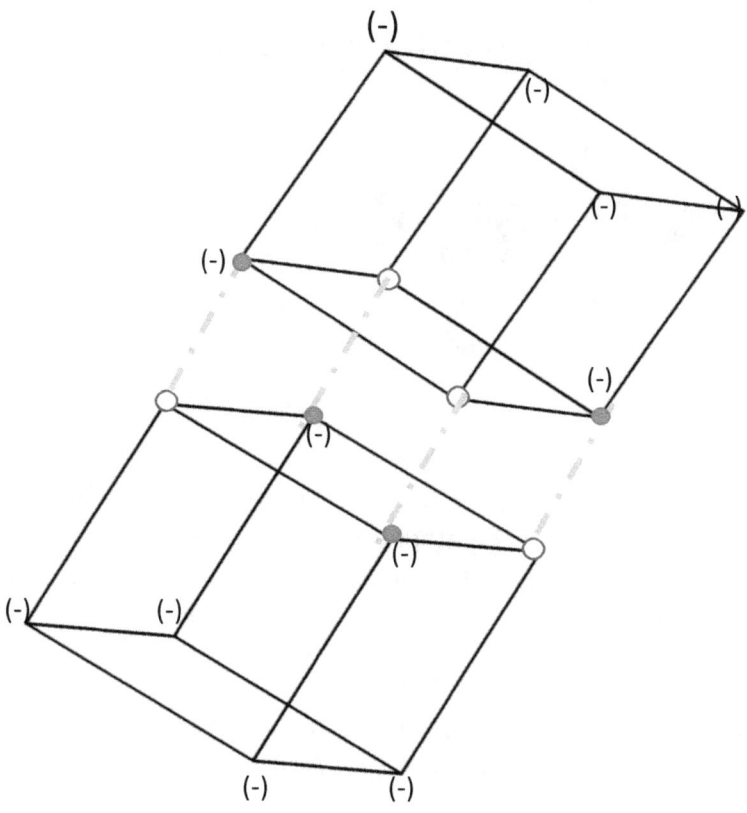

This makes the exterior of an O2 repulsive to other O2 at a distance. Hence, the tendency to become a gas.

This profile of the exterior of each atom, including both the bonding angles, but also adding the positive charge versus negative charge landscape creates the reasoning for better models of bonding.

The traditional ball-and-stick model used only 10x degree angles of a tetrahedron. However, we know that only applies for the right of Shell-2 and Shell-3.

For SO3, Wikipedia has a ball-and-stick model which is really S9O9 because you cannot get 10x degrees to make just one S with 3 Oxygen.

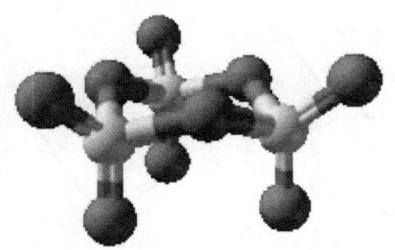

iii

This model is wrong. It expects everything at tetrahedron. It does not consider the magnetics. However, with the electrons in a magnetic field, this model, the positions are much clearer.

The Oxygen are strong receiving molecules which have six of eight electron exterior positions filled. The other two are attractive to other molecules – electrons first.

The 008-O Oxygen has 2m2 and 2c4 as for Shell-2 the differential 'scrunch' is very strong.

The Sulfur also has six exterior electrons, but those sit over a complete Shell 2. That means that the 'scrunch' difference is much less – think 3/2 or about 50% (simplified for this general discussion). The Shell-3 versus Shell-2 at 1/distance-cubed versus 1/distance-squared gives a generally linear change in the ratios.

The secret is that the six electrons can become contributing. Instead of the normal Sulfur -2, in this molecule, the Sulfur is +6, and that makes the bonds 120 degrees and equatorial.

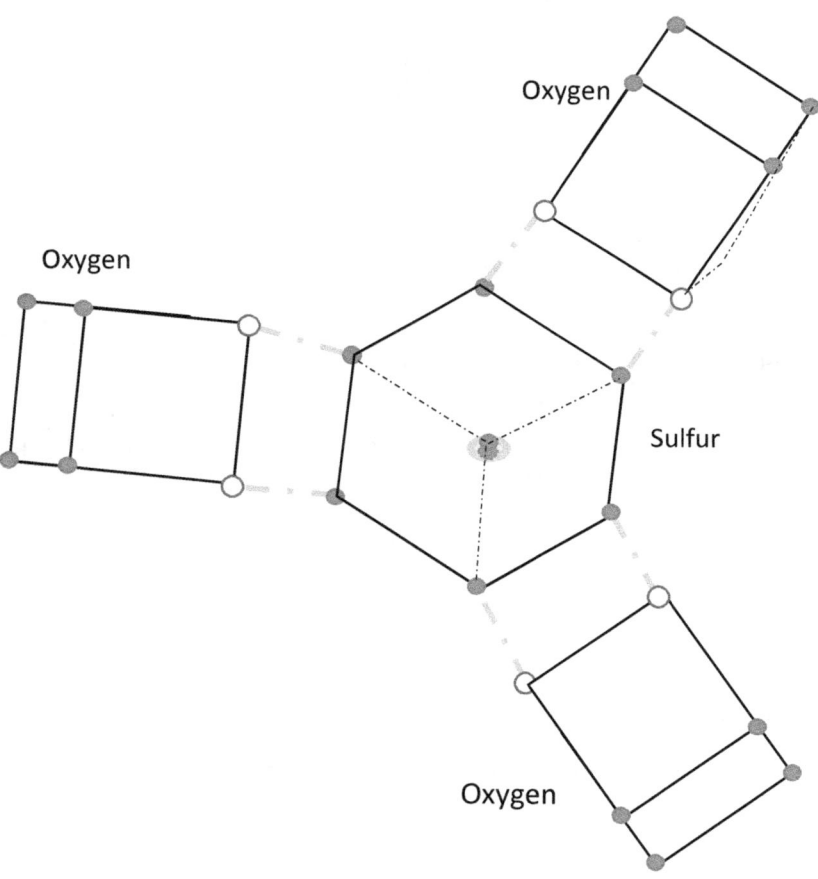

SO3 View from the magnetic poles

From the equator, the electrons move from the normal 2m2 position because of the extra attraction, and they space evenly at the equator. The standard orientation of:

016-S Sulfur – Common -3 Configuration

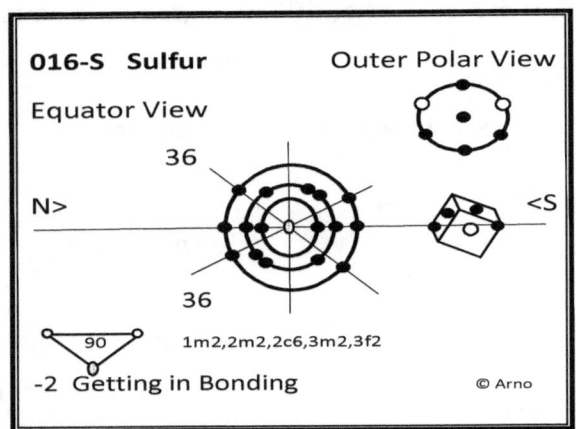

Shell/Subshell/Count	# Electrons	Structure
1m2	2	Magnetic Polar
2m2	2	Magnetic Polar
1c6	6	Scrunched Cube
2m2	2	Magnetic Polar
2f4	4	Faces of Cube

Changes to:

016-S Sulfur – +6 Configuration (SO3)

016-S Sulfur

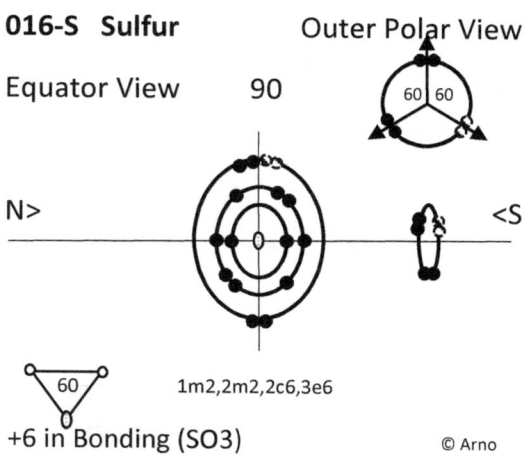

Shell/Subshell/Count	# Electrons	Structure
1m2	2	Magnetic Polar
2m2	2	Magnetic Polar
1c6	6	Scrunched Cube
2m2	0	Magnetic Polar
2e6	6	Equatorial

When bonding with near full halogen atoms, almost every middle-right of the periodic chart atom converts from receiving to contributing. That is from -2 to +6, or -3 to +5.

Further, because of the magnetic strengths and structure of the underlying shell, each of the bond-angle configurations can get determined.

From the side, the Sulfur equator, there are bonds are 120 degrees with all six bonding positions on a plane (equator to the magnetic field of the Sulfur).

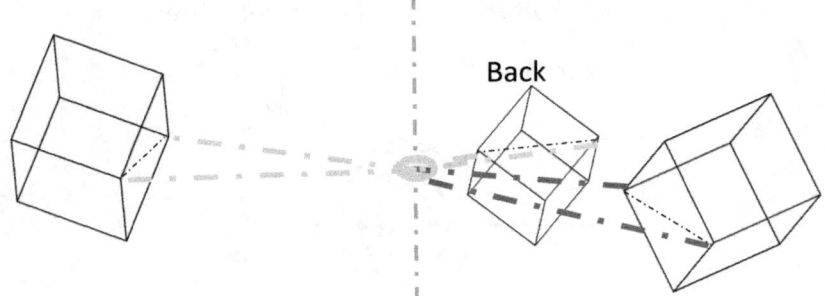

Three 008-O Oxygen given diagonal face bonds with the faces of the underlying atom angled at 90 degrees to the bonds. All are flat.

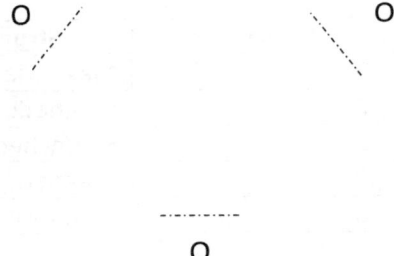

The exterior 008-O Oxygen sit with the diagonal bond position in the Sulfur equator plane, and the face perpendicular to that Sulfur equator.

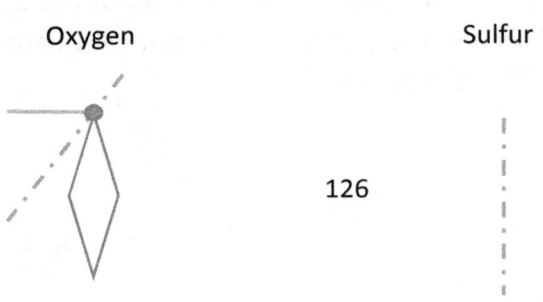

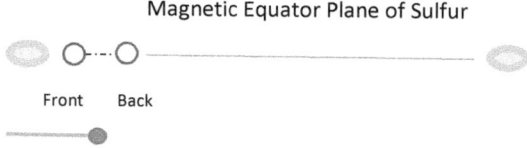

Note that the magnetic orientation of the 008-O Oxygen atoms are different than the Sulfur's, and further the Oxygen atoms might not match each other in magnetic orientation.

Plastics – Polycarbonate Chains

Salts and Ions

Petrochemicals and Energy Release

Endnotes

---

[i] Schrodinger

[ii] The total calculation is complex. It involves:

- Separating the proton and electron calculations to get atom-gravity (called mass)

- Further calculating that atom-gravity for all atoms in a celestial body separately (called an integral in math) to get celestial-body-gravity

- Calculating how far out the electrons in the electron-shells actually sit to do the first calculation cleanly. These resolve certain variances in the measurement of mass already observed.

[iii] https://en.wikipedia.org/wiki/Sulfur_trioxide as of 2017/01/11

www.ingramcontent.com/pod-product-compliance
Lightning Source LLC
Chambersburg PA
CBHW070253230526
45470CB00002B/589